AF882486

AN INTRODUCTORY COURSE IN TOPOLOGY USING A "STUDENT-ORIENTED" APPROACH
1ST EDITION

BY
DR. HORACE SMITH
ASSOCIATE PROFESSOR OF MATHEMATICS
SOUTHERN UNIVERSITY, BATON ROUGE, LOUISIANA

B.S., MISSISSIPPI VALLEY STATE UNIVERSITY;
M.S., Ph.D., UNIVERSITY OF ILLINOIS

INFINITY PUBLISHING

Copyright © 2004 by Dr. Horace Smith

All rights reserved. No part of this book shall be reproduced or transmitted in any form or by any means, electronic, mechanical, magnetic, photographic including photocopying, recording or by any information storage and retrieval system, without prior written permission of the publisher. No patent liability is assumed with respect to the use of the information contained herein. Although every precaution has been taken in the preparation of this book, the publisher and author assume no responsibility for errors or omissions. Neither is any liability assumed for damages resulting from the use of the information contained herein.

ISBN 978-0-7414-1845-6

INFINITY PUBLISHING
1094 New DeHaven Street, Suite 100
West Conshohocken, PA 19428-2713
Toll-free (877) BUY BOOK
Local Phone (610) 941-9999
Fax (610) 941-9959
Info@buybooksontheweb.com
www.buybooksontheweb.com

Preface

This is a very special book. As you read it and the very special style that is used throughout the content, I am sure you will agree with this provocative statement. In the first place, I believe that a one-semester course in point-set topology may be included in an undergraduate mathematics curriculum. I have taught a course like this to undergraduate students for several years. It is designed to offer an honest introduction to point-set topology. To avoid confusion between the mathematical structures and heuristic motivation of the structure, I have employed a B**atz-Beweis style** throughout. Between some definitions and theorems, I have also inserted many brief comments and/or discussions designed to motivate the formal structure. I left many of the proofs to be completed by the reader-although and I have included several proofs for examples.

In my own teaching, I regard the development of facility in constructing proofs as the most important goal of a beginning topology course. Whatever his/her specialty, every mathematician must at some time first cope with an abstract system and learn the mechanics of constructing proofs common to all areas of mathematics.

The instructor exerts considerable indirect influence, often unwitting, not only in the way he sets up the learning environment, but in the details of his verbal and non-verbal interaction with students. The questions he asks, the tone of voice he uses, and his facial expressions all communicate information to the student. For example, in my class I use such phrases as "Show me," "I do not believe that," "Say that again" or "Write that on the black-board."

I do make the claim that the "Student-Oriented Approach" should be used in every mathematics course, nor do I claim that topology is the only course it can be used, I feel that there are other courses in which this approach can and should be used. A few such courses are modern algebra, real analysis and college geometry.

It is important for the instructor to recognize the accepted authority, and for it to be made clear in discussions with students as to the basis on which judgments are made. As far as possible, the discussions and alternate theories should be encouraged, not only to build self-confidence of the students in the power of their own thinking, but also to encourage them to think through the consistencies and inconsistencies of their own ideas. This will encourage tolerance of others.

Instructors must realize that even though they see important reasons for teaching the theories and the students may, for various reasons, want to be acquainted with them, it is not until these theories are assimilated by the student that they are of use to him. To acquire an understanding of theorems and proofs is a long process, and it needs careful though in the preparation of experiences for students.

In the content outline and commentary which follows, I shall describe ways in which each chapter in this instructional material may be covered, along with the reason the chapter has been included

Chapter 0, Preliminary Information and Instructional Treatment, has two purposes: first, make it clear the roll of the student to successful complete this "Student Oriented Approach to topology" using this book; second, to provide a clear guide to the instructor, who wish to use this book and this "Student Oriented Approach"

Chapter 1 serves to introduce the reader to some of the elementary notations of set theory. It not only help to delineate partially the assumptions that are made about set theory and the real numbers, but also provide a easily accessible source of standard definitions and theorems that are not actually part of topology.

Chapter 2 presents a detail study of metric spaces and mapping on metric spaces. This chapter includes many examples of metric spaces.

The main core focus of this book is chapter 3 and chapter 4, which deal with topological spaces. It has been my experience that students are better prepared for later reading on their own if they become acquainted with the variations of "compactness," "separability," etc., that different authors use with varying names. I have made an effort to introduce a reasonable variety of properties and to establish clearly the value of each, as well the exact relationship between the compactness properties, separability properties, etc.

The following observations may be helpful to the instructor who is using this text for the first time and searching for material that can be skipped without any loss of continuity. The instructor may omit some of chapter 1 according to the background of the class.

Acknowledgement. Most of the topologists with whom I have studied, or whose books I have read, have contributed in one way or another to this book. My appreciation goes to my colleagues at Southern University, Baton Rouge, LA. Special thanks go out to my wife, Theola Smith who has given me her love and understanding.

But most of all, to my students go my most heartfelt thanks, without them this book would be very different.

CONTENTS

AND

INSTRUCTIONAL TREATMENT

This is a complete outline of the instructional material on Topology. Included in this section are comments on how to use this material and some suggestions for the instructor. This section should be read in conjunction with step-by-step presentation of the material

This section begins with the assumption that the student does not have a notion of a set of objects. A set is defined (a well defined collection of objects). Then relations on subsets (such as equality and disjointedness) are defined and their properties presented and discussed. These relations are used throughout the student material. The operations of union and intersection are also discussed, and the use of the proofs is emphasized throughout the course.

This section begins with the definition of a collection of subsets of a set X. Properties of families of sets are also discussed.

The Cartesian Product is one of the most important constructions of set theory. In this section the Cartesian Product of N sets, where N is any non-negative integer, is defined and discussed. The students should carefully complete the exercises included at the end of this section.

The concept of a function or mapping is probably one of the most important ideas in all mathematics. The properties of a function are carefully presented and discussed both by the use of symbols and graphs.

At this point the instructor should stress that in order to define a function, three things must be specified: the domain A; a rule assigning a single thing to each element in the domain; and a co-domain B or set that contains the image set of the rule. Here the term "image set" is used to be the set of all members y, in the co-domain such that there exists a member x, in the range with $f(x) = y$. "By abuse of language" uses "the function f" instead of, "the function $f:A \to B$."

The instructor should present various ways of denoting a function (for example, a function may be thought of as an ordered triple as indicated below. Care should always be taken not to use misleading language, however. For example, it is noted that an equation such as $y = f(x)$ is simply the rule of assignment of a function and that is basically incorrect to call an equation a function. To avoid confusing the student, it should be pointed out that the equation $y = f(x)$ may be said to define a function whose domain is the domain of the function denoted by $f:X \to Y$.

The following is the formal definition of a function used in this material: A function is an ordered triple (f, X, Y) such that X and Y are sets and $f \subset X \text{ x } Y$ wit the property: for each $x \in X$, there is one and only one $y \in Y$ satisfying $(x, y) \in f$.

1.5.0 Indexing, Cartesian Products and Projections............34

The concept of a relation is in some sense more general than that of a function. This section considers two types of relations that occur with great frequency in mathematics: equivalence relations and simple ordered relations.

1.6.0 General Cartesian Products................................

The concept of Cartesian Product is extended to the product of a family of sets. The student is required to give several examples of families of sets. Properties and exercises in this section should be discussed in detail.

The idea of continuity is developed out of the definition of a metric space given in Part 2. Since the importance of this section cannot be overemphasized, at this point it may be helpful for the student to recall that definition, and, of course, several examples should also be given. A connection is made between closed sets, open sets, and continuous functions. Homomorphism and one-to-one correspondence are discussed.

Here the ideas of product topology and box topology are defined, and the distinctions between them are drawn. At this point the instructor should make it clear to the student that the product topology is important, as is indicate din this section.

The idea of subspace of a topological space is introduced. The student should be required to generate many subspaces and, by this time should also be able to generate his own examples.

This part discusses some interesting and desirable properties that result when special restrictions are placed on topological spaces. Each of these properties (separated, connected, compact, etc.) should be discussed in detail. Additional properties may be introduced at this point without much confusion for the students.

Here we consider to what extent one can separate closed sets and sets with one element; that is, sets of the form {X}, T_0, T_1, T_2 (Hausdorff Space), T_3, T_4, regular, and normal spaces are defined and discussed. Several unfinished proofs are presented in the student material to help encourage the student to reason independently. The exercises at the end of this section are important and should be discussed.

Here the property of connectedness is discussed in an intuitive way. Several examples and nonexamples are given. In addition to the examples and nonexamples provided, the instructor and the students should offer additional examples and nonexamples.

A formal definition for the term "connected" is given in terms of separated sets. Several properties of connectedness are discussed. Because connectedness is very important, several additional examples of it should be provided and its significance underscored for the students.

As indicated earlier, the instructor who decides to use this material may choose to give proofs to theorems or solutions to the exercises that are different from the ones found in the instructor's material of this study material. In some cases it is suggested that the students write a different proof for each theorem.

It is noted that in some familiar spaces, points may be joined by a graph (path). Some examples of familiar spaces are the real line, and the Euclidean Space R^R. For a Space B and points p and q in B, a path in B from p to q is a continuous map f(a)= p and f(b)= q.

The idea of compactness and several properties of compactness are presented and discussed in this section. The fact that some of these properties are invariant under various kinds of mappings is of fundamental importance in Topology and in the application of Topology to various branches of analysis. Other ideas closely related to compactness may be introduced, if time permits. For example, the notion of sequentially compact spaces may be introduced.

It should be shown that R, R^2, and R^3 are not compact, but each point of these spaces has a compact neighborhood. It should be pointed out and shown in this section, that when a space does have this property, it also possesses certain other useful properties that are true for compact spaces, but not necessarily for all topological spaces.

CHAPTER 0

PRELIMINARY INFORMATION AND INSTRUCTIONAL TREATMENT

This is a very special book; you will find no other book that can be related to the style and approach used throughout the content. This book was written for college topology, which is a course I have taught for several years at Southern University, Baton Rouge. My PhD thesis research contributed a great deal to the style and content of this book. I wanted to make available a course in topology to students with a mathematical background that may not be very strong. I have been dissatisfied for several years with the available texts that may be used for such a course.

In this book, each problem, theorem, lemma, exercise, along with some of the examples not proved or worked out is left for the student to write a proof or solution, and present them in class under the direction, criticism, and supervision of the instructor and other students. To maintain an orderly and continuous coverage of this material, the student should solve each theorem, lemma, and exercise on a rotational basis in alphabetical order, starting on a date set by the instructor.

It is up to the instructor to decide if written examinations are needed. It is the instructor's obligation to clear up proofs or give complete proofs, if the students cannot give such proofs. This material has been developed to give the student a chance to participate; however, it is flexible enough for any instructor to use, even if he does not wish to use this strategy ("Student-Oriented Approach").

In this material, the student and instructor will find that the remarks are very helpful, so they should be read very carefully, as well as other statements. It is the feeling of the writer of this material that a student completing a course of this type ("Student-Oriented" course) will have a better knowledge of topology, and an understanding of proofs of theorems, which will make the study of topology a more meaningful endeavor.

The writer has tried to write the student material in a straightforward style, without unnecessary verbiage or involvement, so that it will be unusually suitable for the students to study with minimum help (proving the theorems and solving the exercises) from the teacher. Important definitions, axioms, and theorems are clearly labeled, and a conscientious effort has been made to utilize each new idea and notation as frequently as possible in order to promote its intelligent use by the student. New material and new points of view are not introduced merely for the sake of novelty, but are brought in only if they make a genuine contribution to the understanding of the concept that can be important to the reader.

The author/instructor preferred to give the students the theorems and exercises without initially providing the proofs or solutions. If a student is unable to readily solve his/her problem on the first attempt, one extension of time is given. If on the second try the student has no valid solution, the problem is then free for any other student in the class to attempt to give a valid solution. If no student can solve the problem, then the instructor gives helpful hints to the students or has them attend a special help session. After a student has correctly completed a proof, he/she is informed as soon as possible. To assure the students of fairness in the process of selection for the order of presentation of problems, whenever problems are scheduled the instructor chooses students by a random process. Students are told that the class will meet four additional days before the presentations are to begin. The author reviews several definitions, solves exercises, gives examples, and proves theorems Before the presentations begins.

The Role of the Student

In the Student-Oriented Approach to teaching Topology, the emphasis is on changing the role of the student from one that is passive to one that is active. Here the word "passive" is used to describe the typical experience for a college student in a mathematics course. The experience is to obtain a textbook at the beginning of the course, and then read one or more sections per lecture. It is rarely the case that another book is consulted or research in the library is done. Most teachers of college mathematics, especially those who teach upper level undergraduate courses and graduate level courses, would agree that the above scenario is not desirable; they would prefer that students read other mathematical material. The word "active" is used to describe the other side of the scenario. These are students who widely read other mathematical literature, take an active part in class activities such as solving problems in class, explaining and giving examples of definitions and theorems, and explaining the details of the theorems. It is clear that college teachers may say, "I have my students do these things. Am I using a student-oriented approach?" The answer in most cases is in the negative, because these are not activities that take place at each class setting, but rather are activities that take place only on special occasions. On the other hand, in a class using the Student-Oriented Approach these are daily activities. The students are encouraged to interact with the instructor and with each other while doing exercises, proofs, and generally the subject matter.

The student has the following specific responsibilities:

1. To write the proof for each theorem and to explain it in class with as little prompting as possible;
2. To write and present satisfactory explanations of examples from the text material;
3. To readily provide additional examples of both definitions and theorems;
4. To solve exercises and write their solutions; and
5. To present and discuss solutions in class.

The Role of the Instructor

The role of the instructor is different in the Student-Oriented Approach than in the conventional lecture approach. The instructor is more a manager of an active learning situation than merely a provider of information. That is in this "different" approach, the instructor must make sure that each student present a clear proof or solution to his assigned exercise. If no student can solve the exercise after several attempts, then it is the instructor's responsibility to solve the exercise and explain the solution to the class. It may also be a need to give individual help sessions to some of the students between class meetings. On the other hand, in the conventional lecture approach, the instructor generally lectures and the students take notes. Therefore, this approach requires that the instructor know his students extremely well so that he may provide each student with the learning experiences that are appropriate at the time.

In the conventional or traditional lecture approach, much use is typically made of examples to present and elaborate upon the concept being taught. Here the word "example" is used in the traditional way; that is, to mean a particular single item, fact, incident, or aspect that is representative of a group or type. However, in Student-Oriented Approach, non-examples are important as well, and are typically used. The word "non-example" is used to mean a particular single item that is not a representative of a particular concept. For example, a non-example of a circle is a square and a non-example of an open interval of the type (4, 7), is a closed interval of the type [1,8]. It should be clear that non-examples are very useful when one wishes to elaborate on concepts. In this setting (SOA) the students are expected to question why a particular object is not a representative of the concept. A plentiful supply of non-examples is needed beforehand so that they can be introduced for discussion at the appropriate moment in class.

In the conventional lecture approach, theorems are typically presented and completely proved by the instructor, thus giving students false impressions about how mathematics is developed. For example, when the instructor writes the proof of a theorem on the board (in a neat and orderly form), the student may think, "This is the way I will do all my proofs." The student may fail to understand that the instructor has had many years of experience. In the Student-Oriented Approach, while some theorems are presented and proved completely, others are not. Typically, the students are encouraged to take part in the development and proof of the details as the proof is developed. In some exercises, the instructor will provide helpful hints that will give the student an indication as to the direction that the proof may take. If a theorem or exercise is not solved in the text, then the students will be required to supply a proof or solution.

The Student-Oriented Approach is much more difficult to use than the conventional lecture approach. But the rewards of the Student-Oriented Approach are very great in terms of the amount of learning that takes place and the attitude toward mathematics that the approach helps to produce.

The Student-Oriented Approach requires the instructor to carefully give each students guidance and help as he/she solves problems and proves theorems, encouraging him to proceed at an individualized pace. The student may need additional examples, further explanations, or exercises to promote understanding. The instructor must continually exercise his own judgment as to what strategies and content that will be most beneficial to the student at his or her particular stage of progress in the course.

During the years in which the author taught topology and developed this approach, he learned the importance of using a variety of means of communication: written and spoken and of using different moods such as jocular, inquisitive, accepting, dour, and so on. For example, in his class, the instructor often uses comments such as the following:

- "Show me. I do not believe that statement."
- "Say that again"
- "Write that down on the board."
- "Are you sure of that?"
- "Can you offer us some evidence of that?"

The author recalls that there was a student who made a statement several times and then was asked by the instructor to write the statement on the board. The student wrote the statement on the board and then said, "Did I say that?" The student was referring to his earlier unsubstantiated remarks.

In using this approach, there are more subtle processes involved here than just the use of changing expressions. Continuous feedback and changes in mood must be used to guide the students to discover proofs and to maintain interest. It is suggested that verbal reinforces be used to develop self-confidence in dealing with abstractions.

The author continuously solicited opinions and questionnaire evaluates from interested persons and students, which provided feedback in the development of the material and the Approach.

When one uses the Student-Oriented Approach, each step of the instructor's explanations must be absolutely clear. The instructor must find many examples (in addition to the ones in the printed material) to help the students understand the concepts, theorems, and exercises. The students should provide additional examples and figures.

The instructor's commentaries are important, as well. Questions that can start a student thinking are one such as the following: "Can this theorem be proved another way?" "Is this theorem still true if we drop this condition?" and "What is the difference between Theorem A and Theorem B?"

It is essential to note that the Student-Oriented Approach to the teaching of Topology develops a deep sense of communication between students and between students and teacher in a short period of time. In just a few class meetings students are getting together for study groups. This is very beneficial, both academically and socially. Students tend to pick up short comments made by the instructor and play them on each other. For example, the author can recall demonstrating a consequence of logical at a class setting. The author requested that each student develop and write a valid proof of a theorem on connected spaces or compact spaces. One student developed but failed to give a valid proof for the theorem: "If $f:X \rightarrow Y$ is not continuous, then $f(X)$ is compact in Y." This was a joke for the class for a few days. This approach to teaching topology requires a great deal of work, but if employed wisely, it can make advanced courses in mathematics not so difficult to teach.

The instructor exerts considerable indirect influence, often unwittingly, not only in the way he sets up the learning environment, but in the details of his verbal and non-verbal interaction with students. The questions he asks, the tone of voice he uses and his facial expressions all communicate information to the students. In the several years that I have taught this topology course I have learned to use all of these wittingly. For example, in my class I use such phrases as "Show-me," "I do not believe that," "Say that again" or "Write that down on the other black-board."

I do not make the claim that the "Student-Oriented Approach" should be used in every mathematics course, nor do I claim that topology is the only course it can be used. I feel that there are other courses in which this approach can be and should be used. A few such courses are modern algebra, real analysis, geometry and I am sure there are others.

It is important for the instructor to recognize accepted authority, and for it to be made clear in discussions with students as to the basis on which judgments are made. As far as possible, the discussions and alternate theories should be encouraged in order, not only to build the self-confidence of the students in the power of their own thinking, but to encourage them to think through the consistencies and inconsistencies of their own ideas and to encourage tolerance of others.

Instructors must realize that even though they see important reasons for teaching the theories, the students may for various reasons, want to be acquainted with them, it is not until these theories are assimilated by the students that they are of use to them. To acquire an understanding of theorems and proofs, whether they are accepted ones or the students' own integration, is a long process, and it needs careful thought in the preparation of experiences for students.

In ordering experiences for students, a further issue has to be taken into account, and that is that the accepted criteria of evaluating their work, which is accepted as logical, may not be that used by the student, especially if the logic the student uses to process the information obtained and come to some belief is different from commonly accepted logic. If the students are to acquire "useful" knowledge, ultimately they have to be trained to accept their own authority and judgment.

In the content and instructional treatment that follows, I shall describe ways in which each chapter in the instructional material may be covered, along with the reason the chapter has been included.

Chapter 1

Introductory Set Theory

It is assumed that what is meant by a set of objects is intuitively clear, and we shall proceed on that basis without analyzing the concept further. At some point of the student's studies he may wish to study set theory more carefully and in greater detail.

1.1.0 Fundamental Concepts

Here we introduce the ideas of set theory and establish basic terminology and notation. Commonly, capital letters; A, B, etc., will denote sets and lower case letters; a, b, etc., will denote the object or elements belonging to these sets. If b belongs to set B, we express this fact by

$b \in B$

If object b does not belong to set B we express this fact by writing

$b \notin B$

The symbol $x = y$ is used throughout this book to mean logical identity and the symbol $x \neq y$ to mean that a member x is different from a member y. If this is the case we write $x \neq y$, also if set A = a set B means that "A" and "B" are symbols for the same set; that is, A and B consist of precisely the same objects.

Subsets

1.1.1 Definition: A set T is a subset of a set S, written $T \subset S$, if and only if for each $x \in T$, $x \in S$. T is a proper subset of S if $T \subset S$ and there is a P $\in$ S with $P \notin T$.

The notation {a, b, c, d} denotes the set consisting of the objects or elements represented by the symbols listed and no other objects. In this text we use the symbols $A \subset B$ and $A \subseteq B$ to mean the same (A is a subset of B). If we wish to emphasize that A is a subset of B and A is not equal to B we will make a clear statement to that fact.

The notation {x: K(x)}, where K(x) is a statement about x, denotes the set of all K's for which K(k) is true and no other objects. This notation is read "the set of all K's such that K(x)." There is nothing special about "x." Thus {x: x is a real number less than 6} and {y: y is a real number less than 6} are the same set of real numbers.

Example 1.1.1 Let S = {1, 2, 3, 4}, T = {2, 3, 4} then $T \subseteq S$. Also, T is a proper subset of S since $1 \in S$ and $1 \notin T$.

Equal Sets

1.1.2 Definition: Two sets S and T are equal, written T = S if and only if S and T have the same members.

1.1.3 Theorem: A set T is equal to a set S if and only if, $T \subseteq S$ and $S \subseteq T$.

Proof: (exercise for the reader)

Example 2.1.1: Let T = {a, b, c, r}, S = {a, b, c} then $S \subseteq T$, but $S \neq T$ since $r \in T$ and $r \notin S$.

Example 1.1.2 Let S = {a, b, c}, T = {c, b, a} then S = T.

Example 1.1.3 Let $S = \{x: x^9 - 5x + 6 = 0\}$, $T = \{x: x \in Z, 1 < x < 4\}$ then S = T.

Remark 1.1.1 Examples 1.1.2 and 1.1.3 should be explained in detail by the students. At this point additional examples would also be useful and helpful to the students. You will find, in this material, claims that are not verified. We do this so the students may give the explanations.

Union and Intersections of Sets

1.1.4 Definition: Let A, and B be two sets, then the union of A and B, written:

$A \cup B = \{x: x \in A \text{ or } x \in B\}$.

The intersection of A and B, written:

$$A \cap B = \{x: \ x \in A \text{ and } x \in B$$

Example 1.1.4 Example: Let $A = \{a, b, c\}$, $B = \{b, c, d, e\}$ then $A \cup B = \{a, b, c, d, e\}$ and $A \cap B = \{b, c\}$.

Empty Set or Null Set

1.1.5 Definition: The empty set written $\varnothing$ or $\{\}$ is the set with no members.

Example 1.1.5: Let $S = \{x: \ x$ is a real number and $x^2 < 0\}$. It is clear that $S = \varnothing$.

Complements

1.1.6 Definition: The complement of set A relative to set B, written:

$$B \setminus A = \{x: \ x \in B \text{ and } x \notin A\}.$$

Example 1.1.6: Let $A = \{a, b, c, d\}$ and $\{a, b, c, e\}$ then:

$B \setminus A = \{e\}$

Remark 1.1.2 for problems dealing with set theory, a pictorial representation of the relations between the sets is frequently helpful. A Venn diagram can give such a picture of representation. For example, a diagram which pictures $A \cap B$ is shown in Figure 1.

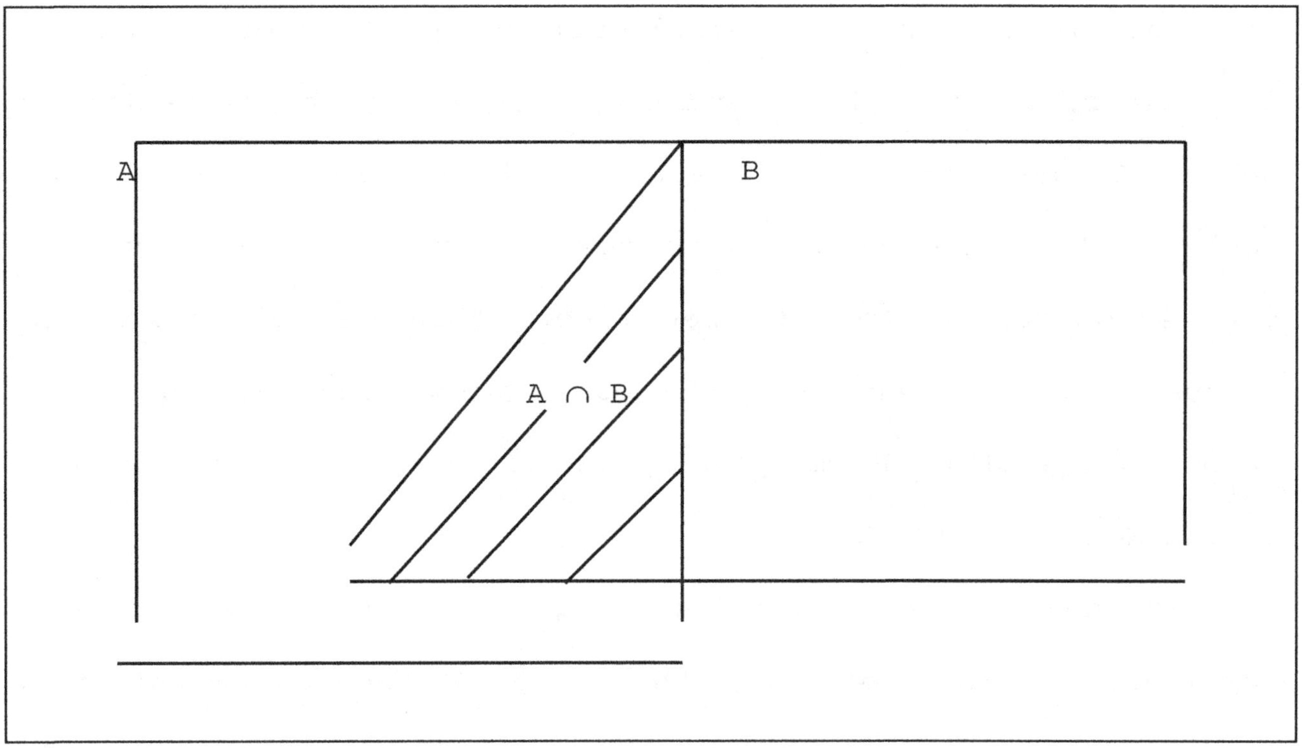

Figure 1

Also, a Venn diagram which pictures T \ S is shown in Figure 2.

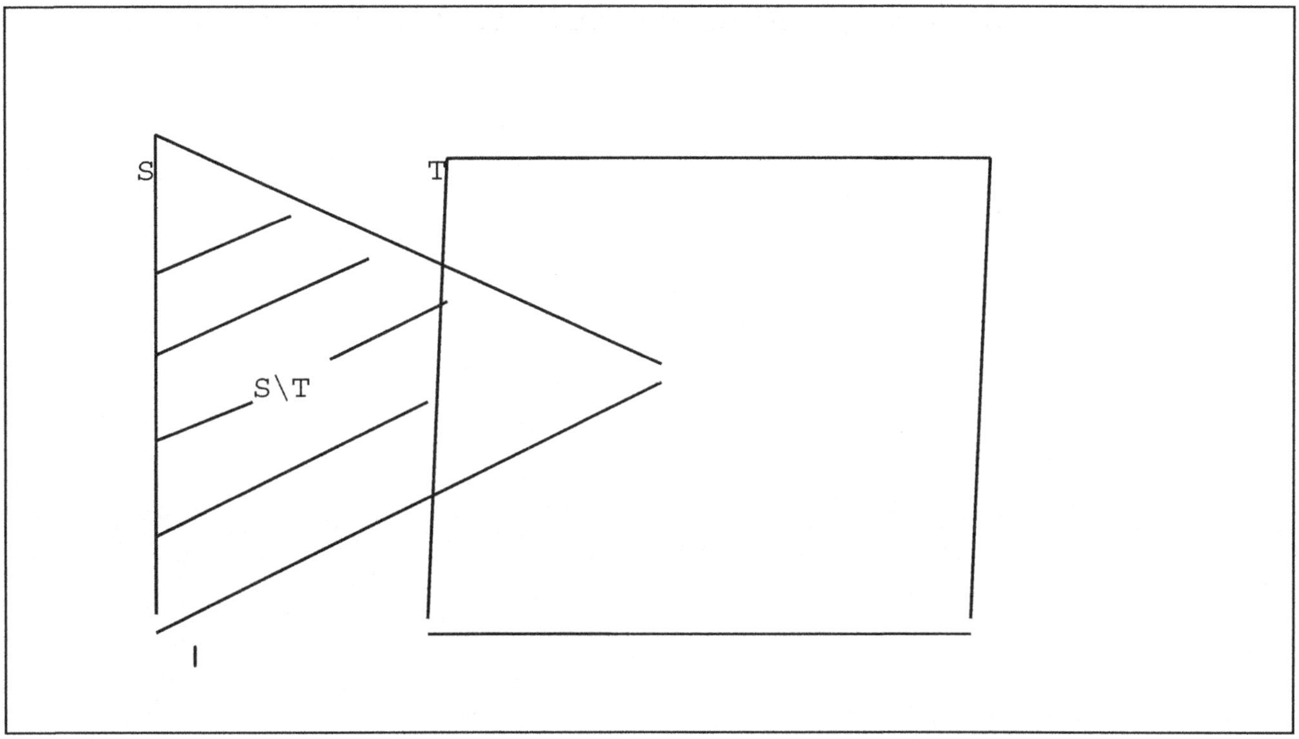

Figure 2

Thus far, we have avoided considering the question of whether or not unions, intersections, and complements of sets are actually sets. To deal with this question and many similar ones that could be found in the pages to come later, it would require a more axiomatic approach to set theory than we are willing to pursue. For an exceptionally readable axiomatic approach to set theory see Paul R. Holmas, *Naive Set Theory*, Princeton, N. J.: D. Van Nostrand, 1960.

It should be noted that the diagram itself does not prove a particular result of set theory. However, the diagram frequently gives some hint as to how a proof can be constructed. This s demonstrated, for example in Figure 3 below, which suggests an approach to the proof that if $A \subseteq B$ and $A \subseteq C$, then $A \subseteq B \cap C$.

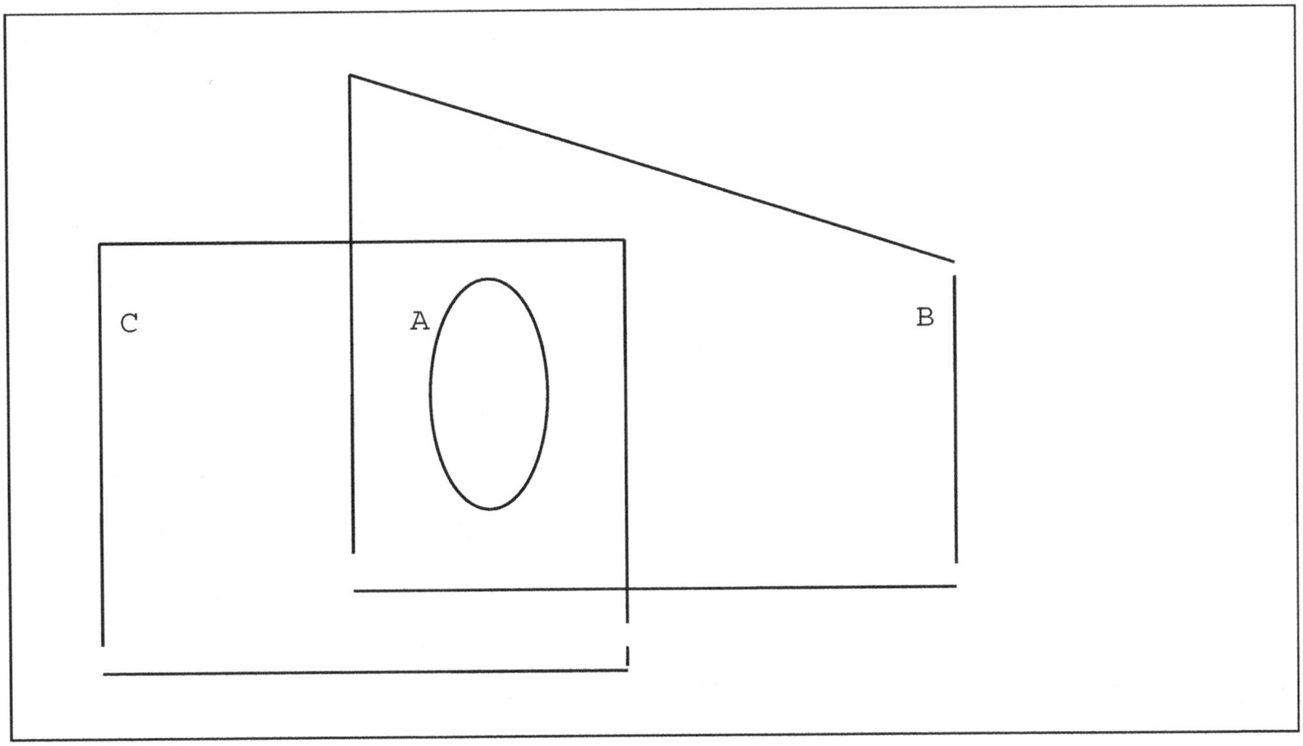

Figure 3

The student should write a proof for the above statement above. Also, $\subseteq$ T and B $\subseteq$ T, then A $\cup$ B $\subseteq$ T (draw a diagram to illustrate this fact).

Disjoint Sets

1.1.7 Definition: Two sets A and B are disjoint if and only if

$$A \cap B = \varnothing.$$

1.1.7 Example: Let s = {x: x is a real number, X > 6), T = {x: x is a real number, x > 4}, then S and T are disjoint, since there is no real number less than 4 and greater than 6, thus S $\cap$ T = $\varnothing$.

1.1.8 Theorem: If A, B, are subsets of X and $A \subset B$, then

$$X \setminus B \subset X \setminus A.$$

Proof: If $X \in X \setminus B$, then $x \in X$ and $x \varnothing B$, since $A \subset B$, since $A \subset X \setminus A$.

1.1.9 Theorem: If $x \setminus B \subset X \setminus A$, then $A \subset B$.

Proof: Suppose $X \setminus B \subset X \setminus A$, then for each $x \in A$ we have that $x \notin X \setminus A$, since $X \setminus B \subset X \setminus A$ we have that $p \in X \setminus B$ therefore, $p \in B$, hence, $A \subset B$.

1.1.10 Theorem: If A and B are sets so that $A \cap B = B$ and

$$A \cup B = B, \text{ then } A = B.$$

Proof: Exercise for the reader

Exercise 1.1.1

In each of the following exercises, all sets are assumed to be subsets of some universal set, X.

1. Prove that if $A \subset B$, A, B, C, sets, then

   a. $A \cap C \subseteq B \cap C$

   b. $A \cap B = A$

   c. $A \cup C \subset B \cup C$

2. Prove each of the following for sets, A, B, and C.

   a. $A \cup \varnothing = A$

b. $A \cap \emptyset = \emptyset$

c. $A \cap A = A \cup A = A$

d. $X \setminus (A \cup B) = (X \setminus A) \cap (X \setminus B)$

e. $X \setminus (A \cap B) = (X \setminus A) \cup (X \setminus B)$

f. $C \subseteq A$ and $C \subseteq B$, then $C \subseteq (A \cap B)$

g. $C \subseteq (A \cap B)$, then $C \subseteq A$ and $C \subseteq B$

h. $A \setminus (A \setminus B) = B$

i. $A \setminus (B \setminus A) = A \setminus B$

j. $A \cap (B \setminus C) = A \cap B) \setminus)A \cap C)$

k. $(A \cap B) \cup (A \setminus B) = A$

1.2.0 Collection of Subsets

1.2.1 Definition: Let $\propto$ be a collection of subsets of X; the union of elements of $\propto$ is defined by

$\cup_{A\in\propto} A = \{p: p \in A$ for at least one $A \in \propto)$, and the intersection of elements of $\propto$ defined by $\cap_{A\in\alpha} A = \{p: p \in A$ for every $A \in \propto\}$.

Note that $\cup_{A\in\alpha} A = \emptyset$ if $\propto = \emptyset$, because it is clear that $\emptyset \subset \cup_{A\in\alpha} A$. Suppose $\cup_{A\in\alpha} A \in \emptyset$ then there is a $p \in \cup_{A\in\alpha} A \not\subset \emptyset = \{p: p \in A$ for at least one $A \in \propto\}$, thus, $\propto \neq \emptyset$, contradiction.

Hence, $A \in \propto \bigcup_{A\in\alpha} A \subset \emptyset$. On the other hand, every p satisfies the defining property for the intersection of elements of $\propto$, that is $\bigcap_{A\in\alpha} A = X$ where $\propto = \emptyset$.

Indexing Set

1.2.2 Definition: Let I and X be sets, if for each $\propto \in I$ we assign a subset $A_\propto$ of X, then we say that the collection of subsets of X, $\{A_\propto: \propto \in I\}$ is indexed by I, and I is called an indexing set for $\{A_\propto: \propto \in I\}$.

Remark 1.2.1 In the case of infinite intersections and unions, some non-intuitive situations can arise:

Example 1.2.1 Let $A_\propto = \{n \in Z: n \geq k\}$ $k = 1, 2, 3, \ldots$ Note that .... $A_3 \subset A_2 \subset A_1$ and $A_n \neq \emptyset$ for each n, yet $\cap A_x = \emptyset$ for if $\bigcap_{x\in Z} A \neq \emptyset$, then $\exists\, p \in \cap n_{x\in Z} A_n \Rightarrow p \in A_n$ for all k, $p \notin A_{p+1} = \{n \in Z: n \geq p + 1\}$ thus $\bigcap_{n\in Z} A_n = \emptyset$.

Example 1.2.2 For each $n \in Z^+$, let $A_n = [0, 1 - 2^{-n}]$ and $B_n = [0, 1 - 3^{-n}]$. Note that each A_n is a proper subset of B_n, and $\bigcup_{n\in Z^+} + A_n = \bigcup_{n\in Z^+} + B_n = [1, 1)$.

1.2.3 Theorem: Let $A_\propto$ be a subset of X for each $\propto \in I$, with $A_\propto \neq \emptyset$ for at least one $\propto \in I$. Then

$$X \setminus \cup_{\propto \in I} A_\propto = \cap_{\propto \in I} (X \setminus A_\propto)$$

Proof. Note that $x \in X \setminus \cup_{\propto \in I} A_2 \Rightarrow x \in X$ and $X \notin \cup_{\propto \in I} A_\propto$

$\Rightarrow x \in X$ and $X \notin A_\propto, \forall_\propto \in I$

$\Rightarrow x \in X \setminus A_\propto, \forall_\propto \in I$

$\Rightarrow x \in X \cap {}_{\propto \in I} (X \setminus A_\propto)$

$\therefore X \setminus \cup_{\propto \in I} A_\propto \subset \cap_{\propto \in I} (X \setminus A_\propto)$. The proof is clearly reversible; hence, we have $\cap_{\propto \in I} (X \setminus A_\propto) \subset x \setminus \cup_{\propto \in I} A$ and, thus, the sets are equal.

1.2.4 Theorem: Let $A_\propto$ be a subset of X for each $\propto \in I$, with $A_\propto \neq \emptyset$ for at least one $\propto \in I$. Then

$$X \setminus \cap_{\propto \in I} A_\propto = \cup_{\propto \in I} (X \setminus A_\propto)$$

Proof: The proof is left as an exercise for the reader.

Exercise Set 1.2

1. Suppose that A, B and C are subsets of the plane, where

$A = \{(x, y): x^2 + y^2 \leq 9\}$,

$B = \{(x, y): x \leq 0 \text{ and } y \geq 0\}$,

$C = \{(x, y): x \leq y\}$. If $\propto$ is the collection of sets $\{A, B, C\}$, graph each of the following sets:

a. $\cap \propto$

b. $\cup \propto$

c. $\cup\propto - \cap \propto$

2. Let $A_n = \{x: x \geq n$, n and x are natural numbers$\}$,

$A = \{A_n$: n is a natural number$\}$. In each of the following, if the statement is true, indicate this fact and justify your answer. If the statement is false, indicate this fact and give a counterexample to justify your answer.

a. $\cup A = A_1$

b. $\cap^n_{k=1} A_k = A_n$

c. $\cap A \neq \varnothing$

3. Let S be a set, and let P(S) denote the collection of all subsets of S; then P(S) is called the power set of S. If S = $\{1, 2\}$, then list the elements of P(S).

1.3.0 Cartesian Product

The Cartesian product is one of the most important constructions of set theory; it enables one to express many concepts in terms of sets. The student should already be familiar with the notion of an ordered pair from analytic geometry. When the Cartesian coordinates (x,y) are used to represent a point in the plane, there are distinct roles played by the 1st and 2nd coordinates: (a,b) = (c,d) if and only if a = c and b = d; to be specific (a,b) = (b,a) if and only if a = b. Thus, one may represent the plane as the collection of all ordered pairs (x,y), where x and y are real numbers. The first (second) element of an ordered pair is called its first (second) coordinate.

Cartesian Product of a Finite Collection of Sets

1.3.1 Definition. (a) Let A and B be two sets, distinct or not. Their Cartesian product, A x B, is the set of all ordered pairs $\{(a,b): a \in A \text{ and } b \in B\}$.

(a) Let X_1, X_2, $X_3 \ldots, X_n$ be sets; then the Cartesian produce of X_1, $X_2, \ldots X_n$, denoted by X_1 x X_2 x X_3 ... x X_n = $\pi^{n}{}_{k=1}$ X_k; is the set $\{(x_1, x_2 \ldots x_n): x_k \in X_k\}$. See Figure 4 below for n = 3.

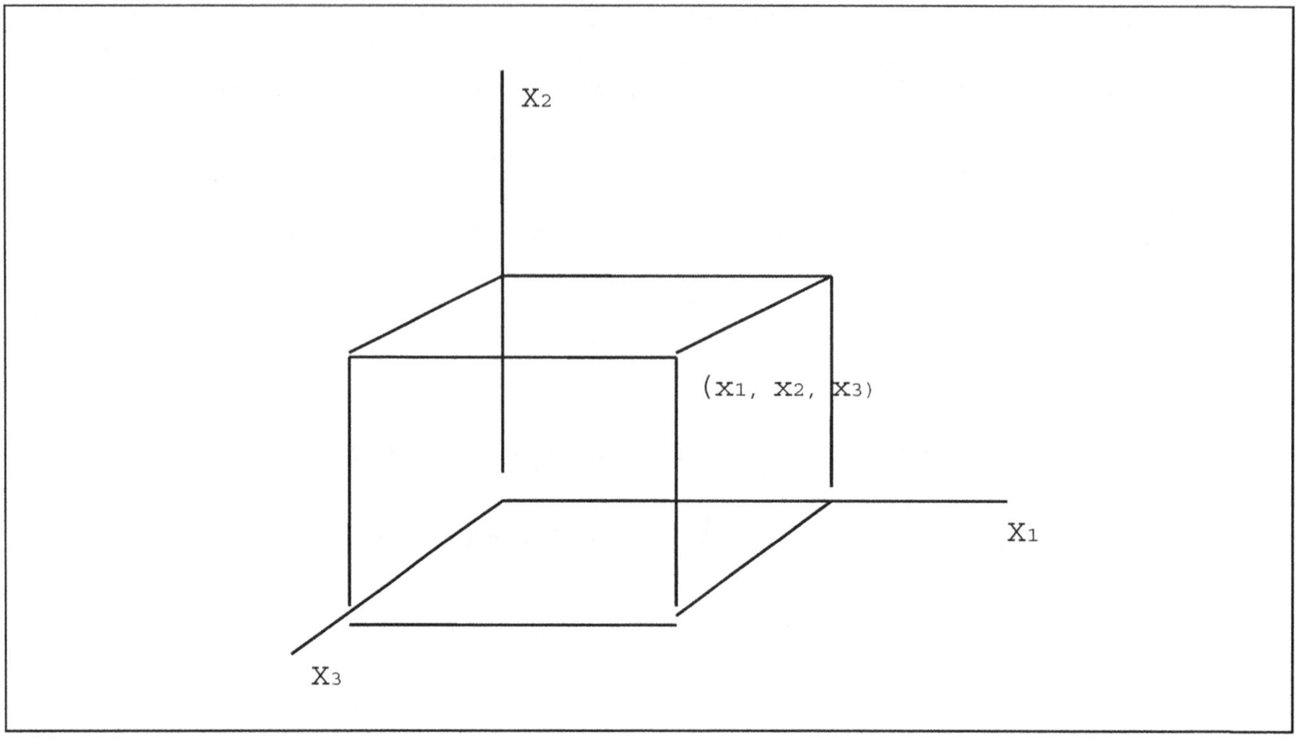

Figure 4

Note that the product of two line segments can be regarded as a rectangle (with inside), the product of a line segment and a circle can be regarded as a (hollow) cylinder. See Figure 5 below.

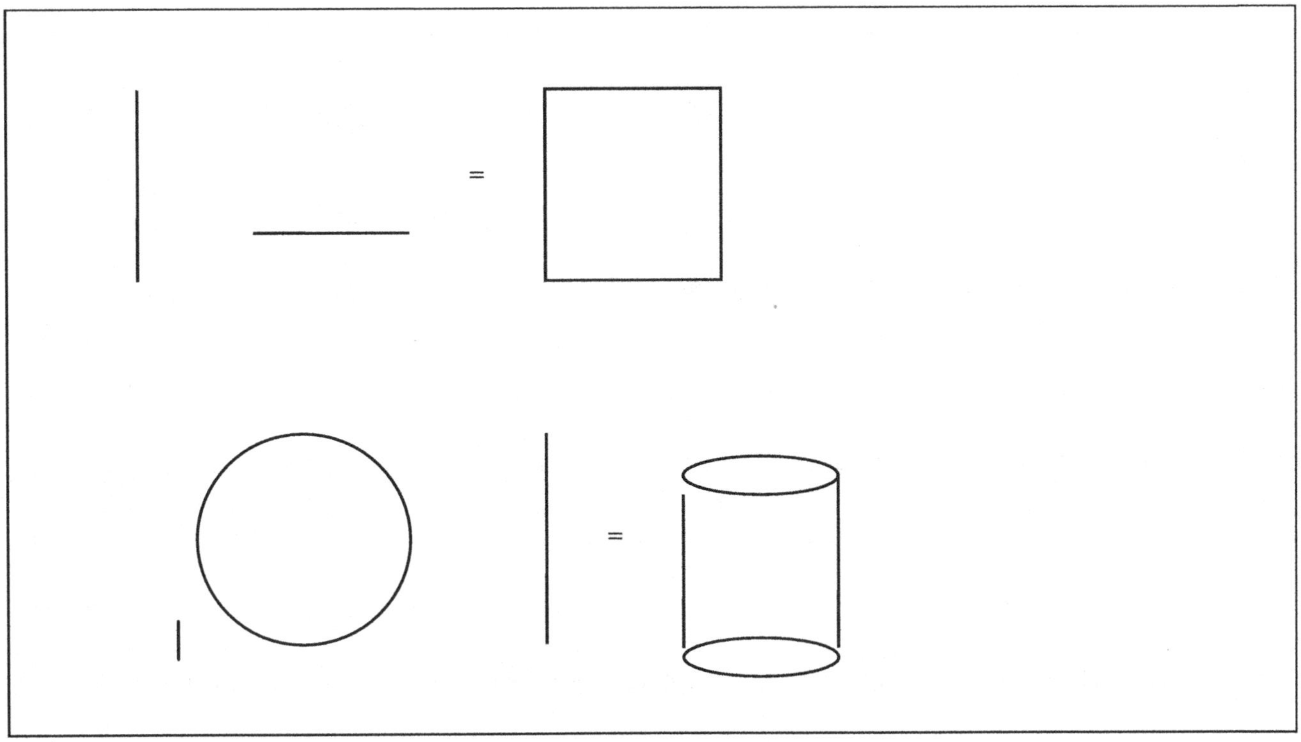

Figure 5

Example 1.3.1 Let A = {1, 3, 5}, and B = {5,9}, then

a. A x B = {1,5), (1,9)(3,5)(3.9), (5,5), (5,9)}

b. B x A = {(5,1), (5,3), (5,5), (9,1)(9,3)(9,5)}

1.3.2 Theorem A x B = ∅ if and only if A = ∅ or B = ∅.

Proof: Supose A x B ≠ ∅ and A = ∅ and B = ∅, then we have(a,b)∈ A x B for at least one pair (a,b), then a ∈ A and b ∈ B, thus A ≠ ∅ and B ≠ ∅.

Conversely, if A ≠ ∅ and B ≠ ∅ and A x B = ∅,we can display some a ∈ A and b ∈ B, then (a,b) ∈ A x B, thus A x B ≠ ∅, contradiction. Hence A = ∅ or B = ∅.

1.3.3 Theorem: If $C \times D \neq \emptyset$, then $C \times D \subset A \times B$ if and only if $C \subset A$ and $D \subset D$.

Proof. The proof is similar to the proof for 1.3.2 and is left as an exercise for the reader.

Remark 4. If a and b are real numbers, the symbol (a,b) is ambiguous, for it may denote either the ordered pair whose first element is a and the second element is b, or the open interval with end points a and b. However, it will usually be clear from the context which meaning to attach to the symbol occurs. It is our feeling that introduction of new Notation will generate more confusion then it will eliminate.

1.3.4 **Theorem**: Let $\{A_\alpha: \propto \in I\}$, $\{B_\beta: \beta \in J\}$ be collections of subsets of X, then $[\cup \{A_\alpha: \propto \in I\}] \cap [\cup\{B_\beta: \beta \in J\}] = \cup \{A_\propto \cap B_\beta: (\propto, \beta) \in I \times J\} \in$

Proof: $C \in \{\cup \{A_\propto: \propto \in I\}] \cap [\cup\{B_\beta: \beta \in J\}]$

$\Leftrightarrow C \in \cup \{A_\propto: \propto \in I\}$ and $C \in \cup \{B_\beta: \beta \in J\}$

$\Leftrightarrow C = A_\propto{}^1$ for some $\propto^1 \in I$ and $C = B_\beta{}^1 \in J$

$\Leftrightarrow C = A_\propto{}^1 \cap B_\beta{}^1$ for some $(\propto^1, \beta^1) \in I \times J$

$\Leftrightarrow C \in \cup \{A_\propto \cap B_\beta: (\propto, \beta) \in I \times J$

$\therefore \{\cup \{A_\propto: \propto \in I\}] \cap \{\cup \{B_\beta: \beta \in J\}] = (A_\propto \cap B_\beta: (\propto, \beta) \in I \times J\}$

Exercise Set 1 - 3.

1. Suppose $A \subset B$ and C is a set. Prove that $A \times C \subset \beta \times C$.

2. Let $A = \{1,2\}$, $B = \{a,b\}$, $C = \{\propto,\beta\}$. List all the members of:

   (a) $A \times (B \cup C)$

   (b) $(A \times B) \cup (A \times C)$

3. Let Z^+ be the set of positive integers, X be the set of real numbers, and for each $\propto \in Z^+$, Let $A_\propto = \{1, 1/2, 1/3, \ldots, 1/\propto\}$. Find

   (a) $\cap \{A_\propto: \propto \in Z^+\}$

   (b) $\cup \{A_\propto: \propto \in Z^+\}$

   (c) $\cap \{X \setminus A_\propto: \propto \in Z^+\}$

   (d) $\cup \{X \setminus A_\propto: \propto \in Z^+\}$

4. If $A = \{x \in R: 1 \leq x \leq 2\}$ R the set of real numbers and B $= \{y \in R: 3 < y < 4\}$ then graph

   (a) $A \times B$

   (b) $B \times A$.

1.4.0 Functions

The notations of functions or mapping is basic in all mathematics. it is probably one of the most important ideas in all mathematics. Fortunately, it is a simple idea: A function from a set X to a set Y is a rule that associates one and only one element of Y with each element of X: it is an unambiguous (see Figure 6) way to go from set X to Set Y.

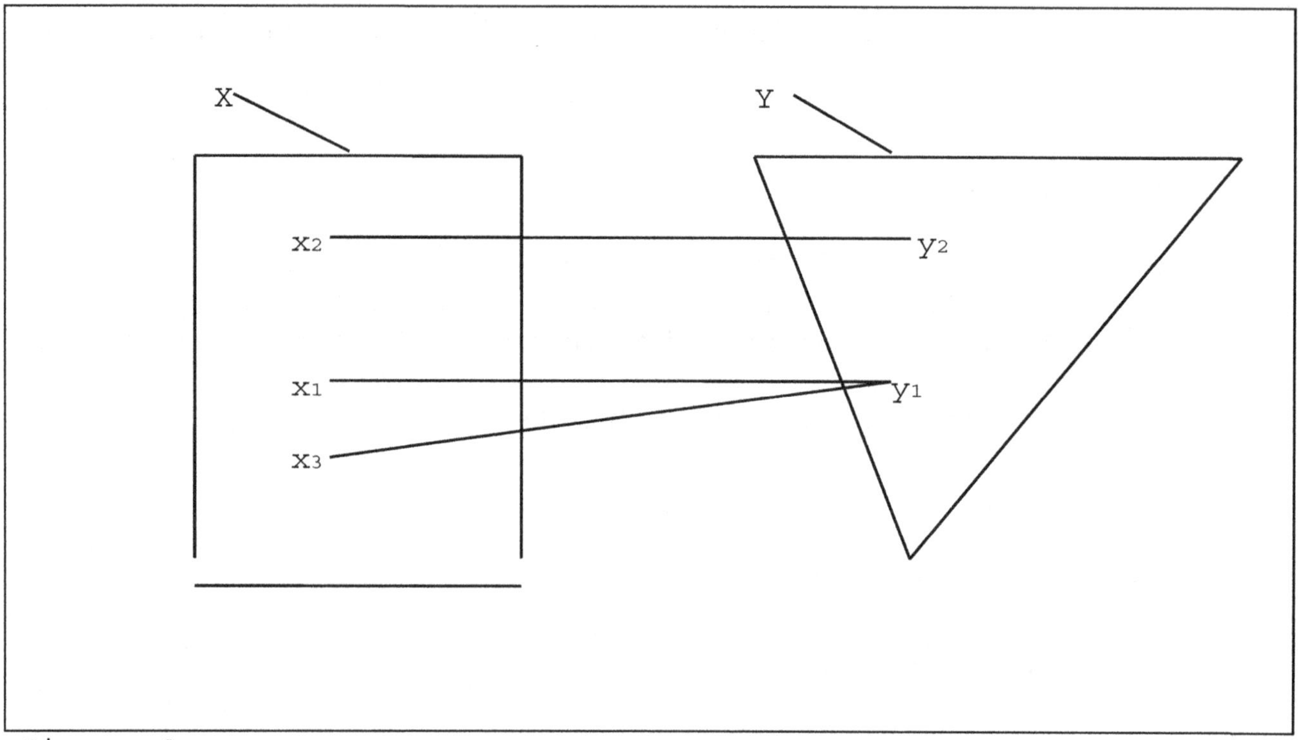

Figure 6

Function

1.4.1 Definition. Let X and Y be two sets. A function f from X to Y (written $f:X \to Y$) is a subset of X x Y so that whenever (x_1,y_1) and (x_1,y_2) both belong to the set of f then $y_1 = y_2$.

Remark 1.4.1 This definition may be confusing at first, but a little thought will convince you that it simply makes the intuitive idea of a function mathematically precise. Note that we do not prohibit x_1 and x_2 from being paired with the same y, even if $x_1 \neq x_2$.

Example 1.4.1 Let X = {a, b, c}, y = {1, 2, 3} and define f(a) = 1, f(b) = 2 = f(c); then f is a map from X to Y.

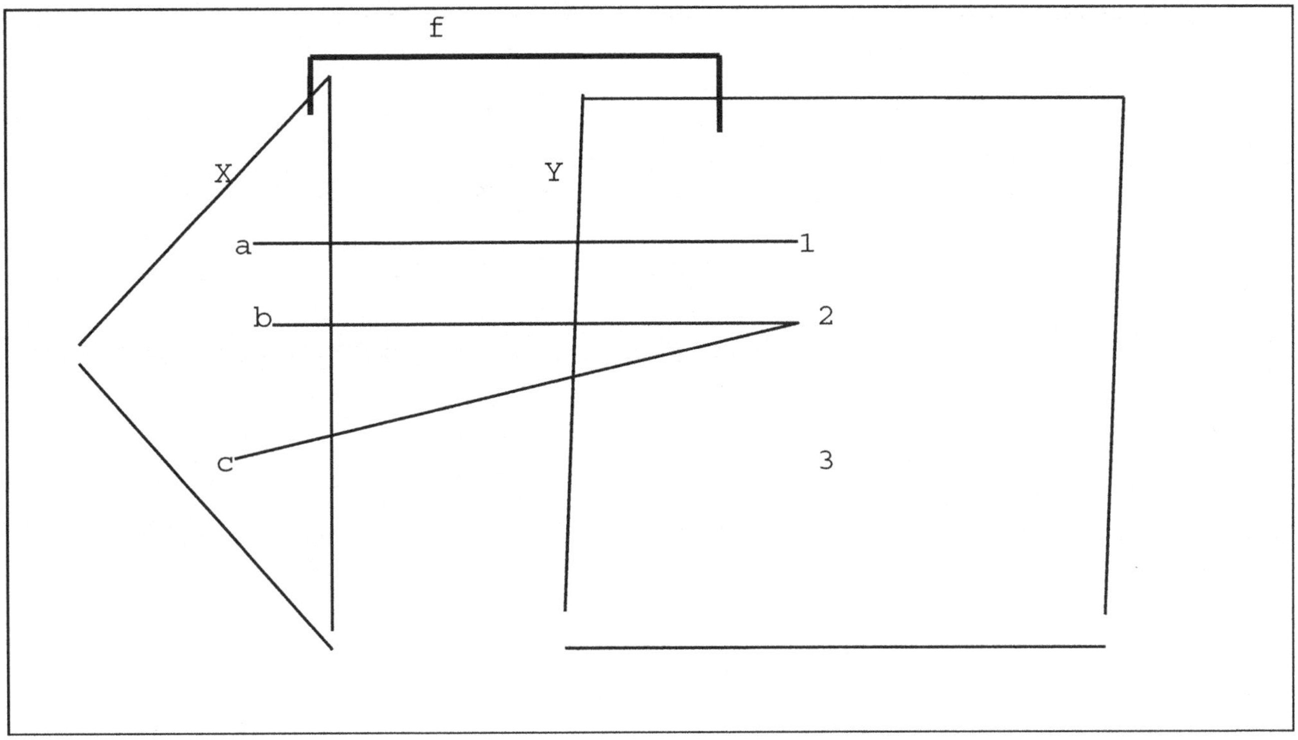

Figure 7

Domain and Range

1.4.2 Definition. Let $f:X \to Y$ be a map (function) then the domain of f denoted by Dom f = X. The range of f denoted by range f - $\{y \in Y: (x,y) \in f$: for at least one $x \in X\}$.

Remark 6. It should be clear that in example 1.4.1, Dom f = {a, b, c} = X and range f = {1, 2}; when we write $f:X \to Y$ we will mean that Dom f = X and range of f = $f(x) \subset Y$.

Image and Inverse Image

1.4.3 Definition. Let $f:X \to Y$ be a function then:

(a) For each $A \subset X$, $F(A) = \{f(x): x \in A\} \subset Y$ is called the image of A in Y under f.

(b) For each $B \subset Y$, $f^{-1}(B) = \{x \in X: f(x) \in B\} \subset X$ is called the inverse image of B in X under f.

Example 1.3.2. Let $f:R \to R$, where R the set of real numbers, be the function so that $\forall x \in R$.

$f(x) = x^2$ also

Let A = [-1,1], B = [0, 1/9] then f(A) = [0,1] and $f^1(B)$ = [-1/3, 1/3].

1.4.4 Theorem: Let $f:X \to Y$ be a map and $\{B_\alpha\}_{\alpha \in I}$ be a collection of subsets of Y then

(a) $f^{-1}(\cup_{\alpha \in I} B_\alpha) = \cup_{\alpha \in I} f^{-1}(B_\alpha)$

(b) $f^{-1}(\cap_{\alpha \in I} B_\alpha) = \cap_{\alpha \in I} f^{-1}(B_\alpha)$

(c) $f^{-1}(B_\alpha \setminus B_\beta) = f^{-1}(B_\alpha) \setminus f^{-1}(B_\beta)$

Proof. We establish (b) only; the proof for (a) and (c) are similar and are left as exercises for the reader.

$x \in f^{-1}(\cap_{\alpha \in I} B_\alpha) \Leftrightarrow f(x) \cap_{\alpha \in I} B_\alpha$

$\Leftrightarrow f(x) \in B_\alpha, \forall_\alpha \in I$

$\Leftrightarrow X \in f^{-1}(B_\alpha), \forall_\alpha \in I$

$\Leftrightarrow X \in \cap f^{-1}(B_\alpha), \forall \alpha \in I$

Hence, we have that $f^{-1}(\cap_{\alpha \in I} B_\alpha) \subset \cap_{\alpha \in I} f^{-1}(B_\alpha)$ (by$\Rightarrow$) and $\cap_{\alpha \in I} f^{-1}(B_\alpha) \subset f^{-1} \cap_{\alpha \in I} B_\alpha)$ (by $\Leftarrow$) thus $f^{-1}(\cap_{\alpha \in I} B_\alpha) = \cap_{\alpha \in I} f^{-1}(B_\alpha)$.

Composition of Functions

Suppose one wanted to consider the function f defined by F(x) = Cos(Sin (x)) for x a real number. This function is actually complicated. However, since the sine and cosine are well-known functions, it is not difficult to study f by means of what we already know about sine and cosine. This technique of studying complicated functions by means of composition of simpler ones has proved to be an extremely useful technique at all levels of mathematics. This reader who has studied elementary calculus will, for example, recall the technique of finding the derivative of a complicated function by means of the chain rule for composite functions.

Composition of Two Functions

1.4.5 Definition. Let f and g be two maps so that f: $X \to Y$ and $g: Y \to Z$; then their composition written g o $f: X \to Z$ is defined as the map g o $f(x) = g(f(x))$, $x \in X$ (see Figure 9 below).

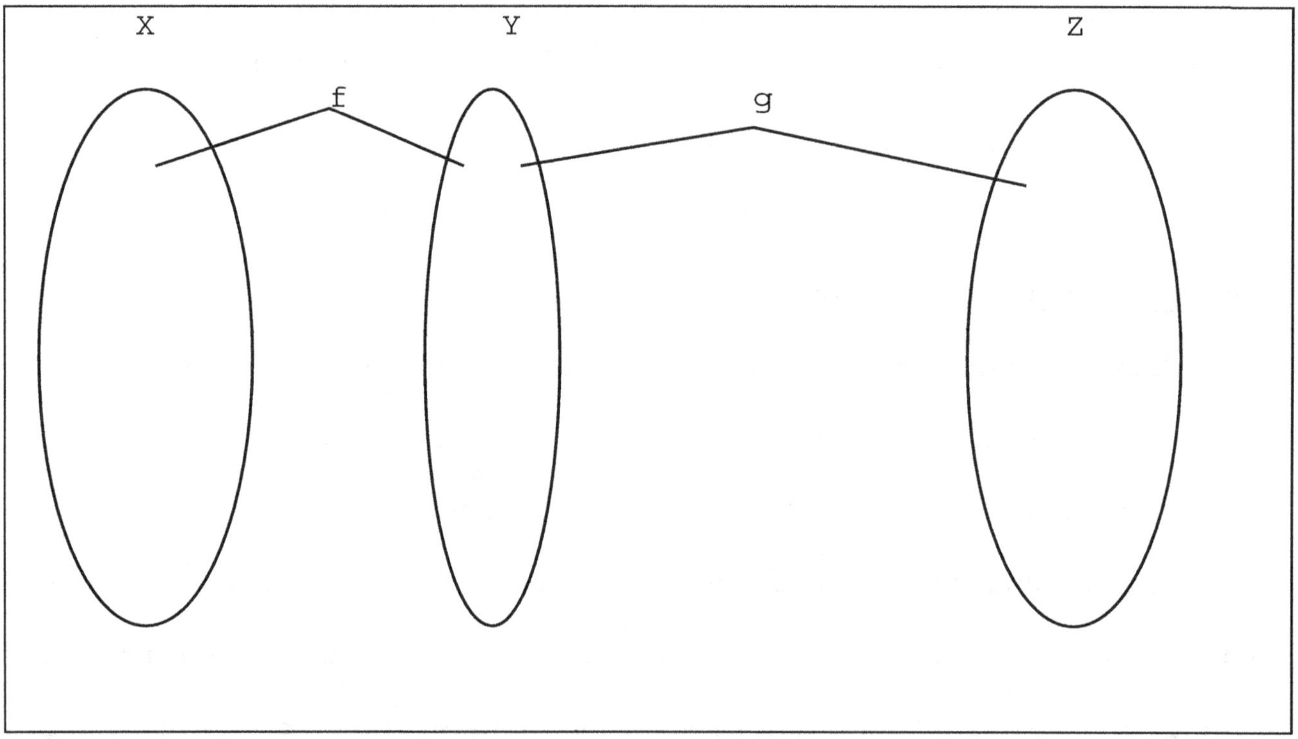

Figure 8

1.4.6 Theorem. Let f and g be maps so that $f: K \to Y$ and $g: Y \to Z$; then $(g \circ f)^{-1} = F^{-1} \circ g^{-1}$.

Proof: $x \in (g \circ f)^{-1}(Z)$

if and only if $(g \circ f)(x) = g(f(x)) \in Z$

if and only if $f(x) \in g^{-1}(Z)$

if and only if $X \in f^{-1}(f(x)) \subset f^{-1}(\{g^{-1}(Z)\})$

if and only if $x \in f^{-1} o\ g^{-1}(Z)$,

hence, $(g \circ f)^{-1} = f^{-1} \circ g^{-1}$.

1.4.7 Theorem: Let f, g, and h be functions such that the range of f is a subset of the domain of g and the range of g is a subset of the domain of h. Then h o (g o f) = (h o g) o f.

Proof: By hypothesis h o (g o f) and (h o g) o f are well defined. Also, Dom f = Dom h o (g o f) = (h o g) o f. Also, for each X Dom f we have that (h o g) o f)(x) = h o g(f(x)) = h(g(f(x))), or (h o (g o f))(x) = h o g(f(x)) - h(g)f(x))). Thus (h o g) o f = h o (g o f). (See Figure 10)

1.4.8 Definition: Consider the map $f:X \to Y$ then

(a) f is 1-1 (or injective) if whenever) $f(x_1) = f(x_2)$, then $x_1 = x_2$

(b) f is onto (or surjective if $f(X) = Y$

(c) f is bijective if f is both injective and surjective.

Remarks 1.4.1 Let $f:Z \to Z$, Z the set of integers defined by $f(x) = 2x$ and $g:Z \to Z^+ \cup \{0\}$ defined by $g(x) = x^2$, then f is one-to-one but not onto, g is onto but not 1-1.

1.4.9 Definition: A set S is said to be finite if it is empty or if there is a bijective correspondence

$$f:S \longrightarrow \{1,\ldots\ \ldots\ \ldots,n\}$$

for some $n \in Z_+$. In the case, of the empty set we say that S has 0 elements; in all other finite cases we say that S has n elements.

1.4.10 Theorem: Let n be a positive integer. Let A be a set; let a_0 be an element of A. If there exist a bijective map f of the set A with the set $\{1,\ldots\ \ldots, n+1\}$, then there exist a bijective map g of the set A-$\{a_0\}$ with the set $\{1,\ldots\ \ldots\ \ldots,n\}$.

Proof: Assume there is a bijective map

$$f:A \longrightarrow \{1,\ldots\ \ldots\ \ldots,n+1).$$

If f maps a_0 to the number n+1, things are easy; in that case, the map $g = f:A-\{a_0\}$ is the desired bijective map of $A-\{a_0\}$ to $\{1,\ldots\ \ldots\ \ldots,n+1)$. Otherwise, let $f(a_0) - m$, and let a_1 be the point of A such that $f(a_1) = n+1$. Then $a_1 \neq a_0$. Define a new map

$$h:A:\longrightarrow \{1,\ldots\ \ldots\ \ldots,n+1\}$$

by setting

$$h(a_0) = n+1,$$

$$h(a_1) = m,$$

$$h(x) = f(x) \text{ for } x \in A - \{a_0\}.$$

It is easy to check that h is a bijection.

Now we are back to the easy case g = h:A − $\{a_0\}$ is the desired bijection of A − $\{a_0\}$ to $\{1, \ldots \ldots \ldots, n+1\}$.

1.4.11 Theorem: Let n be a positive integer. Let A be a set; let a_0 be an element of A. If there exist a bijective map g of the set A −$\{a_0\}$ with the set $\{1, \ldots \ldots \ldots, n\}$. Then there exist a bijective map f of the set A with the set $\{1, \ldots \ldots, n+1\}$.

Proof: Assume that there is a bijective map

$$G: A - \{a_0\} \longrightarrow \{1, \ldots \ldots \ldots, n\}.$$

We then define a map $f: A \longrightarrow \{1, \ldots \ldots \ldots, n+1\}$. By setting

$$f(x) = g(x) \text{ for } x \in A - \{a_0\}.$$

$$f(a_0) = n+1.$$

It is clear that f is a bijective map.

1.4.12. Theorem: If S is finite, there is no bijective map of S with a proper subset of itself.

Proof: Exercise for the reader.

1.4.13. Theorem: If B is a subset of the finite set A, then B is finite.

Proof: Exercise for the reader.

1.4.14. Theorem: If B is a proper subset of A, then the number of elements of B is less than the number of elements of A.

Proof: Exercise for the reader.

1.4.15 Definition: A set S is said to be infinite if it is not finite.

1.4.16 Definition: A set S is said to be countable infinite if there is a bijective corresponfence

$$f: Z_+ \to S,\ Z_+ \text{ the set of positive integers.}$$

1.4.17 Example: Let Z be the set of all integers, then Z is countable infinite. One may check this easily by the function

$$f(n) = 2n \text{ if } n > 0 \text{ and}$$

$$f(n) = 1 - 2n \text{ if } n < 0$$

to see that f is a bijection.

1.4.18 Theorem: A subset of a countable set is countable.

Proof: Exercise for the reader.

1.4.19 Theorem: The set $Z_+ \times Z_+$ is countable infinite.

Proof: In view of Theorem 1.4.17, it is suffice to construct an injective map from $Z_+ \times Z_+$ to z_+ to do this we define

$$f: Z_+ \times Z_+ \to z_+ \text{ by}$$

$$f(m,n) = 2^m 3^m,$$

it is clear that f is injective. For suppose $2^m 3^m = 2^p 3^q$. If $n < p$, then $3^m = 2^{p-n} 3^q$, contradicting yhe fact that 3^m is oddd for all m. Therefore $p = n$. as a result, $3^m = 3^q$. Then if $m < q$ we have that $1 = 3^{q-m}$, another contradiction. Hence $m = q$. Hence the proof is complete, that is $Z_+ \times Z_+$ is countable infinite.

1.4.20 Theorem: The set Q_+ of all positive rational numbers is countably infinite.

Proof: Exercise for the reader.

1.4.21 Theorem: The set Q of all rational numbers is countably infinite.

Proof: Exercise for the reader.

1.4.22 Definition: A subset S of the real numbers is said to be **inductive** if for each $x \in S$ then the number $x + 1 \in S$.

1.4.23 Theorem: The set of all positive integers Z_+ is inductive.

Proof: Exercise for the reader.

1.4.24 Teeorem: Let β be the collection of all inductive subsets of R (the set of real numbers) that contains 1 ,

Z_+, the set the set of all positive integers,

then

$$Z_+ = \cap_{a \in \beta} A.$$

Proof: Exercise for the reader.

1.4.25 Theorem: If $n \in Z_+$, then every subset $\{1, \ldots \ldots n\}$ has a smallest element.

Proof: If $n \in Z_+$, we use $\{1, \ldots \ldots n\}$ to denote the set $\{x: x \in Z_+ \text{ and } 1 \le x \le n\}$, since there is no integer y between n and n + 1 we have

$$\{1, \ldots \ldots \ldots n, n+1\} = \{1, \ldots \ldots \ldots n\} \cup n + 1.$$

Let Z_α be the set of all positive integers n for which this statement is true. Then Z_α contains 1, since if n + 1, the only subset of $\{1, \ldots \ldots \ldots n\}$ is the set $\{1\}$ itself. Then, supposing Z_α contains n, we now show that it contains n + 1. So let S be a nonempty subset of the set $\{1, \ldots \ldots \ldots n\}$. If S consists of the single element n + 1, then that element is the smallest of S. Otherwise, consider the set $S \cap \{1, \ldots \ldots \ldots n\}$, which is nonempty. Because $n \in Z_\alpha$, this se has a smallest element, which will automatically be the smallest element of S also. Thus Z_α is inductive and contains 1, so we conclude that $Z_\alpha = Z_+$. Thus the statement is true for $n \in Z_+$.

1.4.26 Theorem: (Well-ordering property). Every nonempty subset of Z_+ has a smallest element.

Proof: Exercise for the reader.

1.4.27 Definition: Let S be any ordered set, then x is a lower bound for S if and only if for each $p \in S$ we have that $x \leq p$.

1.4.28 Definition: Let S be any ordered set, then y is an upper bound for S if and only if for each p ∈ S we have that p ≤ y.

1.4.29 Definition: Let S be any ordered set, then x is the greatest lower bound for S provided x is a lower bound for S and for each p ≥ x, then p is not a lower bound for S.

1.4.30 Definition: Let S be any ordered set, then y is the least upper bound for S provided y is an upper bound for S and for each p ≤ y, then p is not an upper bound for S.

1.4.29 Definition: Let S be any ordered set, then x is the smallest (minimum) element in S provided x ∈ S and x is the greatest lower bound for S.

1.4.30 Definition: Let S be any ordered set, then y is the largest (maximum) element in S provided $y \in S$ and y is the least upper bound for S.

1.4.31 Theorem. Show that any set of real that has an upper bound has a least upper bound.

Proof: Exercise for the reader.

1.4.31 Theorem. Show that any set of real that has a lower bound has a greatest lower bound.

Proof: Exercise for the reader.

Exercises Set 1-4.

1. State precisely what is meant when a map $f:X \to Y$ is not 1-1, not onto.

2. If map $f:X \to Y$ then show that:

   (a) $f(A \cup B) = f(A) \cup f(B)$ for subsets A and B of X.

   (b) $f(A \cap B) \subset f(A) \cap f(B)$ for subsets A and B of X.

   (c) $f(\cup_{\alpha \in I} A_\alpha) = \cup_{\alpha \in I} f(A_\alpha)$, $A_\alpha \subset X$, $\forall_\alpha \in I$

   (d) $f(\cap_{\alpha \in I} A_\alpha) \subset \cap_{\alpha \in I} f(A_\alpha)$, $A_\alpha \subset X$, $\forall_\alpha \in I$

(e) Give an example to show that $\subset$ cannot be replaced by = in part (b).

3. If map $f:X \to Y$, then:

(a) For each $A \subset X$, $A \subset f^{-1}[f(A)]$

(b) Give a condition that would make $A = f^{-1}(f(A))$ for each $A \subset X$.

4. Let $f:X \to Y$ and $g:Y \to Z$ be maps

(a) If f and g are injective, show that $g \circ f$ is injective.

(b) If $g \circ f$ is injective, what can be said about injectivity of f and g?

(c) If f and g are surjective, show that $g \circ f$ is surjective.

(d) If $g \circ f$ is surjective, what can be said about surjectivity of f and g.

(e) Summarize your answers to (a)-(d) in the form of a theorem.

5. Make a list of all injective maps

$$f:\{1,\ldots\ldots 3\} \to \{1,\ldots\ldots 3,4\}$$

6. How many injective maps

$$f:\{1,\ldots\ldots 8\} \to \{1,\ldots\ldots 8,9\}?$$

7. Show that if B is not finite and $B \subseteq A$, then A is not finite.

8. Show that the set of real numbers $(\infty,\infty) = \{x: \infty < x < \infty\}$ has no greatest lower bound or least upper bound.

9. Show that the set of real numbers $(a,b) = \{x: a < x < b\}$ has a greatest lower bound and a least upper bound.

10. Show that $3Z_+ \times 3Z_+$ is countable infinite.

11. Show that the set of real numbers $(a,b) = \{x: a < x < b\}$ has no greatest or smallest element.

Remarks 1.4.2 Let us at this point discuss an axiom from set theory. The Axiom of Choice is frequently used in topology, as well as other areas of mathematics. Without mentioning it, we have already used this axiom, or a consequence of it, several times in our proofs. In a formal system of mathematics, we realize that we do need an axiom to guarantee the existence of a set A.

Axiom of Choice: Given any nonempty family $(A_\propto: \propto \in J\}$ of nonempty pair wise disjoint sets there exist a set S consisting of exactly one element from each set.

1.5.0 Indexing, Cartesian Products and Projections

1.5.1 Definition. (a) Let J be an indexing set for the family $S = \{A_i: i \in J\}$, Then J is an indexing set for S.

(b) Let T = a: $a = \pi a_i$, ai $\in A_i$ for each i $\in$ J]. Then T is the Cartesian product of members of S.

(c) Let $\propto \in J$, the map $P_\propto$ of $\pi_{i \in J} A_i$ onto $A_\propto$ defined by $P_\propto(x) = x_\propto$ for each $x \in \pi_{i \in J} A_i$ is called the $\propto$th projection.

Exercises Set 1-5.

1. If each A_i, of the family $\{A_i: i \in J]$ has exactly one element. Show that $\pi_{i \in J} A_i$ has one element.
2. If $J = \varnothing$, then $\pi_{i \in J} A_i$ has exactly one element. What is this element.

3. If $J \neq \emptyset$ and for some $A_i \in (A_i: i \in J; A_i = \emptyset$, then $\pi_{i \in J} A_i = \emptyset$.

4. Let $[A_i: i \in J]$ be a nonempty family of nonempty sets; then $P_i: \pi_{i \in J} A_i \to A_i$ is surjective.

5. If $A_i = \{1,2\}$, $A_2 = \{4\}$, $A_3 = \{3,5\}$ $\emptyset$, then what is $\pi^3_{i=1} A_i$? What is $P_1: \pi^3_{i=1} A_i$? What is $P_3: \pi^3_{i=1} A_i$?

6. If $B_\propto \subset X_\propto$ for each $\propto$, then what is $P_\propto^{-1}(B_\propto)$? If $b_\propto \in B_\propto$, then what is $P_\propto^{-1}(B_\propto)$?

7. In $\pi_{\propto \in J} X_\propto$, if $B_\propto \subset X_\propto$ for each $\propto \in J$, show that for each $\propto \in J$. $\pi_{\propto \in J} X_i \times B_\alpha = \cap_{\propto \in J} P_\propto^{-1} (B_\propto)$ (Hint: use the result of exercise 5.

Chapter 2

Metric Spaces

2.1.0 Introduction

Thus far, we have largely restricted our study to abstract sets and functions or relations on abstract sets. The study generally becomes more useful when a structure of some kind or another is imposed on the set to be studied. Recall that by the absolute value of a real number, written $|x|$ we mean x if $x \geq 0$ and -x if $x < 0$. Using the absolute value function, we can introduce the notion of the distance between two points x,y in the real number set R, written

$d(x, y) = |x-y|$.

Note that each distance function has the following properties:

For each $x,y,z \in R$

a. $d(x, y) = 0$ if and only if $x = y$

b. $d(x, y) = d(y, x)$

c. $d(x, y) + d(y, z) \geq d(x, z)$.

R^n for n = 1,2 or 3 and arbitrary sets, also we shall study some properties of metric spaces.

2.2.0 Metric Space

Definition, Examples, and Properties of Metric Spaces

2.1.1 Definition. Let X be a nonempty set of real numbers and $d:X \times X \to R$ be a function that satisfies the following for each x, y, $z \in R$:

(a) $d(x, y) \geq 0$

(b) $d(x, y) = 0$ if and only if $x = y$

(c) $d(x, y) = d(y, x)$

(d) $d(x, z) \leq d(x, y) \div (d(y, z)$

then (x, d) is a metric space, and d is a metric for x.

Remarks. 2.1.1 In fact for any set X it is always the case that $d:X \times X \to R$ defined by

$$d(x, y) = \begin{cases} 1 \text{ if } x \neq y \\ 0 \text{ if } x = y \end{cases}$$

for each x, $y \in X$. Then (X, d) is a metric space. It is a simple verification that with this definition of d, and set X, (X, d) is a metric space.

Example 2.1.1 Let $X = R^2$ and let $d:X \times X \to R$ defined by $d(x, y) = [(x_1 - y_1)^2 + (x_2 - y_2)^2]^{1/2}$, x, $y \in X \times X$, then (X, d) is a metric space. The vader should verify this fact.

2.1.2 Theorem: Let (X, d) be a metric space, and let $d^*: X \times X \to R$ be defined by $d^*(x, y) = K(d(x, y))$ where K is a real number greater than zero, then (X, d^*) is a metric space.

Proof. (Exercise for the reader).

2.1.3 Theorem: Let $d: X \times X \to R$ so that for each $x, y, z \in X$

(a) $d(x, y) = 0 \Leftrightarrow x = y$

(b) $d(x, y) \leq (x, z) + d(y, z)$.

Then (X, d) is a metric space.

Proof. By (b) $d(x, y) = d(x, z) + d(y, z)$ for each $x, y, z \in X$. Take $z = x$ then $d(x, y) \leq d(x, x) + d(y, x)$

$= 0 + d(y, x)$

$= d(y, x)$.

Also, we have $d(y, x) \leq d(y, y) + d(x, y)$ by (b).

$= 0 + d(x, y)$

$= d(x, y)$

Hence, we have $d(x, y) = d(y, x)$.

Suppose $d(x, y) < 0$ for some $x, y \in X$ now by (a) $0 = d(x, x)$ and by (b) we have $0 + d(x, x) \leq d(x, y) + d(x, y) = 2\, d(x, y) < 0$ contradiction hence $d(x, y) \geq 0$.

Clearly, $d(x, y) \leq d(x, z) + d(z, y)$ since by (b) $d(x, y) \leq d(x, z) + d(y, z)$ and we have shown that $d(y, z) = d(z, y)$ Thus we have $d(x, y) \leq d(x, z) + d(y, z) = d(x, z) + d(z, y)$.

Proof. (Exercise for the reader).

Exercise Set 2-1.

1. Let (X, d) be a metric space. Set $d'(x, y) = \dfrac{d(x,y)}{1 + d(x,y)}$ for each $x, y \in X$. Prove that (X, d') is a metric space.

2. Let (X, d) be the metric space and Set $d'(x, y) = d(0, x) + d(0, y)$ for each $x, y \in X$, $x \neq y$. If $x = y$ set $d'(x, y) = 0$. Prove that (X, d') is a metric space.

3. Let $X = \{X_i: I = 1, 2, ...\}$. Set $d(x_n, x_m) = 1 + \dfrac{1}{n+m}$ for $n \neq m$ and $d(x, y) = 0$ if $x = y$. Prove that (X, d) is a metric space.

4. Let $d: R^+ \times R^+ \to R$ be defined by $d(x, y) = |1/x - 1/y|$ where R^+ are the positive real numbers, then (R^+, d) is a metric space.

5. Let $X = R^2$ and let $d: X \times X \to R$ be defined by $d(x, y) = |y_1 - y_1| + |x_2 - y_2|$, where $x = (x_1, x_2)$ and $y = (y_1, y_2)$. Show that (X, d) is a metric space.

2.3.0 Continuity

In calculus, the first occurrence of the word "continuity" is with respect to function $f:R \rightarrow R$ where R is the set of real numbers. At this point let us review the definition of continuity from calculus.

Continuity from Calculus

2.3.1 Definition. Let $f:R \rightarrow R$ be a function; then f is continuous at a point $c \in R$, if for each real number $\in > 0$, there is a real number $\delta > 0$ such that

$|f(x) = f(c)| < \in$ whenever $|x - c) < \in$.

The function f is continuous if it is continuous at each of its points.

Remark. 2.3.1 Because the definition of continuity was initially formulated in terms o the phrase "degree of closeness," we can easily devise a definition of "continuity" applicable to metric spaces, since we need only use the distance functions of these metric spaces to measure "degree of closeness."

Continuity and Metric Spaces

2.3.2 Definition. Let (X, d), (y, d*) be metric spaces and let $c \in X$. A function $f: X \to Y$ is said to be continuous at the point $c \in X$ if for each $\in > 0$, there exist $\delta > 0$, and $\in > 0$, δ, $\in$ are real numbers, such that $d^*(f(x), f(c)) < \in$ whenever $d(x, c) < \delta$.

The function $f:X \to Y$ is said to be continuous if it is continuous at each point of X.

Remark 3.2.2 Definitions such as those above are given to create or formulate precise characterizations of what we feel are significant properties. There are, in a certain sense, tests that we may apply to see whether or not they do so. There are certain functions that we "know" are "continuous" that is, we know that they possess this property that we are trying to characterize. This type of testing of a definition takes the form of proving theorems. If it should turn out that a function we know to be continuous is not continuous in accordance with these definitions, then, although these definitions may be precise, they would not furnish a precise characterization of the property we have in mind when we say a function is continuous.

2.3.3 Theorem: Let (X, d) and (Y, d*) be metric spaces. Let $f:X\to Y$ be a constant function; then f is continuous.

Proof. Let a point $c \in X$ and $\epsilon > 0$, a real number be given. Choose a real number $\delta = \epsilon$. then whenever $d(x, c) < \delta$ we have that $d^*(f(x), f(c)) = d^*(k, k) = 0 < \epsilon$.

2.3.4 Theorem: Let (X, d) be a metric space. then the identity function $i:X\to X$ is continuous.

Proof. (Exercise for the reader).

Remark 2.3.3 There is one situation we should consider for which the notation $f:X\to Y$ that we have adopted for a function from a metric space (X, d) into a metric space (Y, d*) is ambiguous. Consider the metric space (X, d) and (X, d*) with the same underlying set. If we simply write $f:X\to X$ for a function, it is impossible to tell which metric space is denoted by the first occurrence of X and which is denoted by the second. For this reason we write $f:(X, d)\to(X, d^*)$ and not $f:X\to X$ as a function from the metric space (X, d) into the metric space (X, d*).

2.3.5 Theorem: Let $id:(R^3, d)\rightarrow(R^3, d^*)$ be the identity function. Then id is continuous, where the distance functions

$$d(x, y) = \text{Maximum } \{d(x_k, y_k,): 1 \le K \le 3\}$$

$$d^*(x, y) = \left(\Sigma_{k=1}^{3} (x_k - y_k)^2\right)^{½} \quad \text{where}$$

$x = (x_1, x_2, x_3) \in R^3$, $y = (y_1, y_2, y_3) \in R^3$.

Proof. Let $x = (x_1, x_2, x_3) \in R^3$. Let $\in > 0$ be given. Choose $\delta = \in/3$. Suppose $y = (y_1, y_2, y_3)$ is such that $d(x, y) < \delta$; that is Maximum $\{|x_k - y_k|: 1 \le k \le 3\}$. Then $d^*(id(x),id(y))$ =

$$d^*(x, y) = \left(\Sigma_{k=1}^{3} (x_k - y_k)^2\right)^{½} < (3\delta^2)^{½} = (\in^2)^{½} \ge \in$$

Therefore, for any real number $\in > 0$, there is a $\delta - \in/3$, such that we have a $d^*(id(x), id(y)) < \in$ whenever $d(x, y) < \delta$. Hence the id function is continuous.

Remark 2.3.4 One elementary theorem about continuous functions is the fact that the composition of a finite collection of continuous functions is again a continuous. The following theorem establishes this fact for two continuous functions, which are sufficient for a finite collection.

2.3.6 Theorem: Let (X_1, d_1), (X_2, d_2), and (X_3, d_3) be metric spaces. Let $f_1{:}X_1 \to X_2$ and $f_2{:}X_2 \to X_3$ be continuous with $c \in X_1$ and $f_1(c) \in X_2$, respectively, then $f_2 \circ f_1{:}X_1 \to X_3$ is continuous at $c \in X$.

Proof. Let $\in > 0$ be given. We must find a $\delta > 0$ so that whenever $x \in X_1$ and $d_1(x, c) < \delta_1$, then $d_3(f_2(f_1(x)), f_2(f_1(c))) < \in$ (see Figure 1 below).

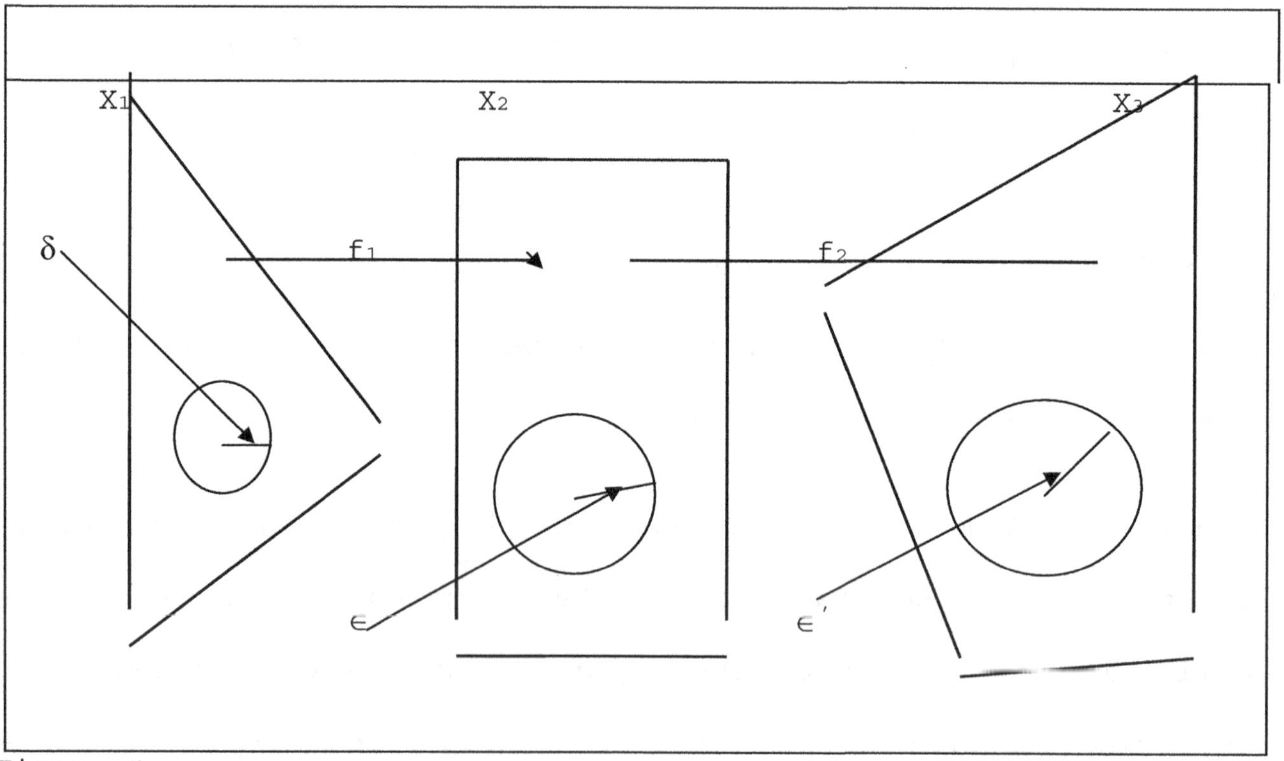

Figure 1

Since F_2 is continuous at $f_1(c)$, there is an $\in' > 0$, such that whenever $y \in X_2$ and $d_2(y, f_1(c)) < \in'$, then $d_3(f_2(y), f_2(f_1(c))) < \in$. Now using the fact that f_1 is continuous at c, we know that

$\epsilon' > 0$, there is a $\delta > 0$, such that $x \in X_1$ and $d_1(x,c) < \delta$ imply that $d_2(f_1(x), f_1(c)) < \epsilon'$ and hence, we have $d_3(f_2(f_1(x)), f_2(f_1(c))) < \epsilon$. Thus, we have $f_2 \circ f_1 : X_1 \to X_3$ is continuous at $c \in X_1$.

2.3.8 Corollary. Let (X_1, d_1), (X_2, d_2) and (X_3, d_3) be metric spaces. Let $f_1 : X_1 \to X_2$ and $f_2 : X_2 \to X_3$ be continuous, then $f_1 \circ f_2\ X_1 \to X_3$ is continuous.

Proof. (Exercise for the student).

2.4.0. Open Ball and Neighborhoods

In our definition of continuity of a function f at a point c in a metric space (X,d), we are concerned with how f transforms those points $\delta \in X$ such that $d(x,c) < \epsilon$. To make this clear let us review the idea of open sets of R^n. Recall that a subset U or R^n is said to be open in R^n provided that for each $p \in U$, there is an $\epsilon > 0$ such that $B(p; \epsilon) = \{ x: d(x,p) < \epsilon\} \subset U$. We now formulate the definition of open balls in a metric space.

Open Ball

2.4.1 Definition. Let (X,d) be a metric space. Let $p \in X$ and $\delta > 0$ be given. The set $U \subset X$ where $U = \{x \in X: d(p,x) < \delta\}$ is called the open ball about p of radius d and is denoted by B(p; d) (see Figure 2 below).

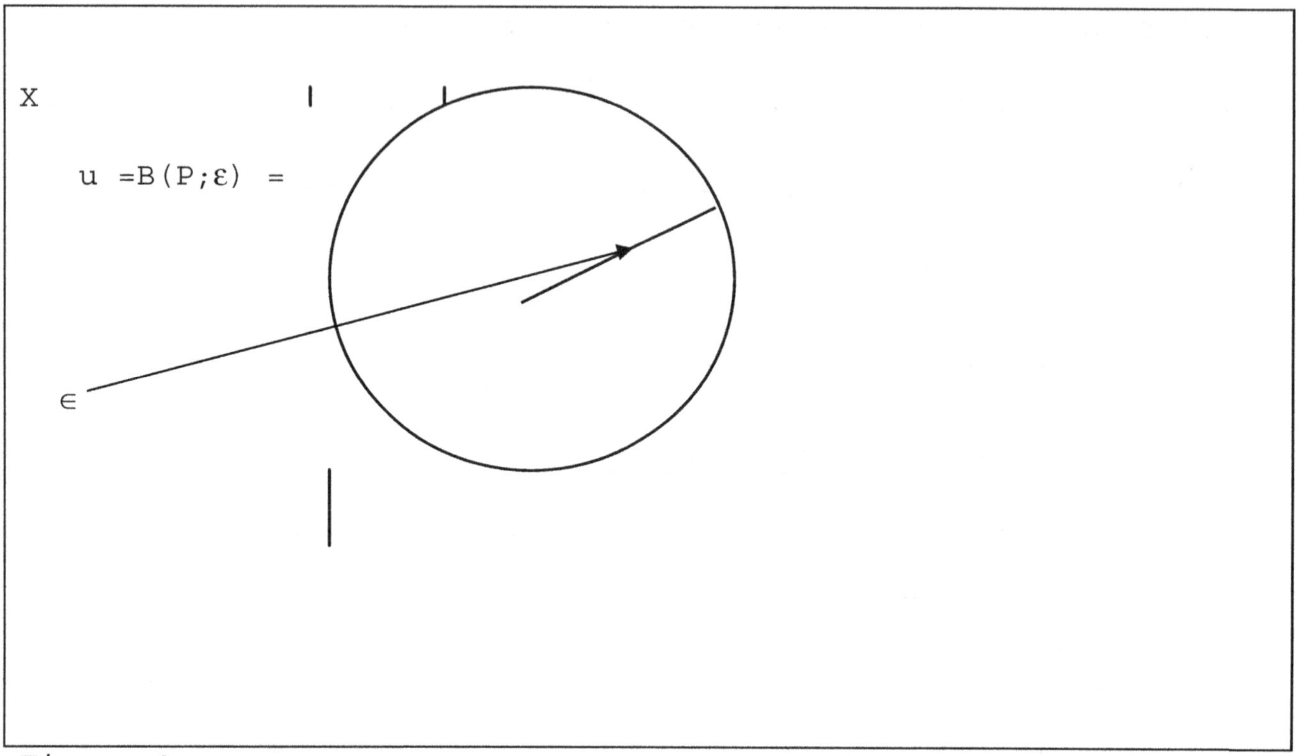

Figure 2

Remarks 2.4.1 Clearly, this definition of open ball is in agreement with our definition of open sets in Rn. Consider B(p: $\in/2) \subset B(p;\in)$ for any $\in > 0$. Similarly, if (X, δ) and (Y,d') are metric spaces and $f:X \rightarrow Y$, we have $y \in B_{d'}(f(p); \subset)$ if and only if $d'(y,f(p)) < \in$. Thus, we may prove the following theorem.

2.4.2 Theorem: A function $f:(X,d) \rightarrow (Y,d')$ is continuous at a point $P \in x$ if and only if for each $\in > 0$

there is a $\delta > 0$ such that $f(B(p;\delta)) \subset B(f(p); \in))$.

Proof: ($\Rightarrow$) $y \in f(B(P;\delta))$ then there exists an $x \in B(P;\delta)$ with $f(x) = y$, but $x \in B(p;\delta)$ we have $d(x,p) < \delta$, since f is continuous at p we have that $d(f(x),f(p)) < \in$ thus $y = f(x) \in B(f(p); \in)$. Hence $f(B(p;\delta)) \subset B(f(p); \in))$.

($\Leftarrow$) Exercise for the reader (see Figure 3 below).

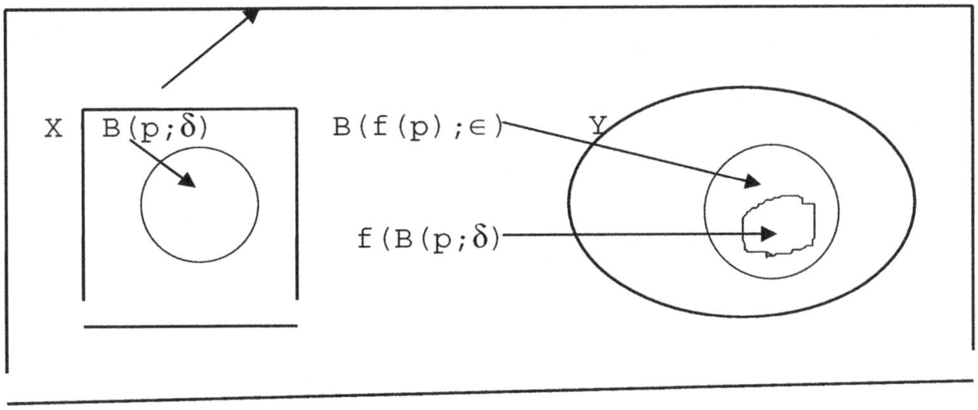

Figure 3

2.4.3 Theorem: A function $f:(x,d)\rightarrow(y,d')$ is continuous at a point $p \in X$ if and only if given $\in > 0$ there is a $\delta > 0$ such that $B(p;\delta) \subset f^{-1}(B(f(p); \in))$.

Proof. Exercise for the reader. (Hint: Use the fact that if $f:X \rightarrow Y$ is a map and $f(A) \subset B$, then $A \subset f^{-1}(B)$ with $A \subset X$ and $B \subset Y$) See Figure 4 below.

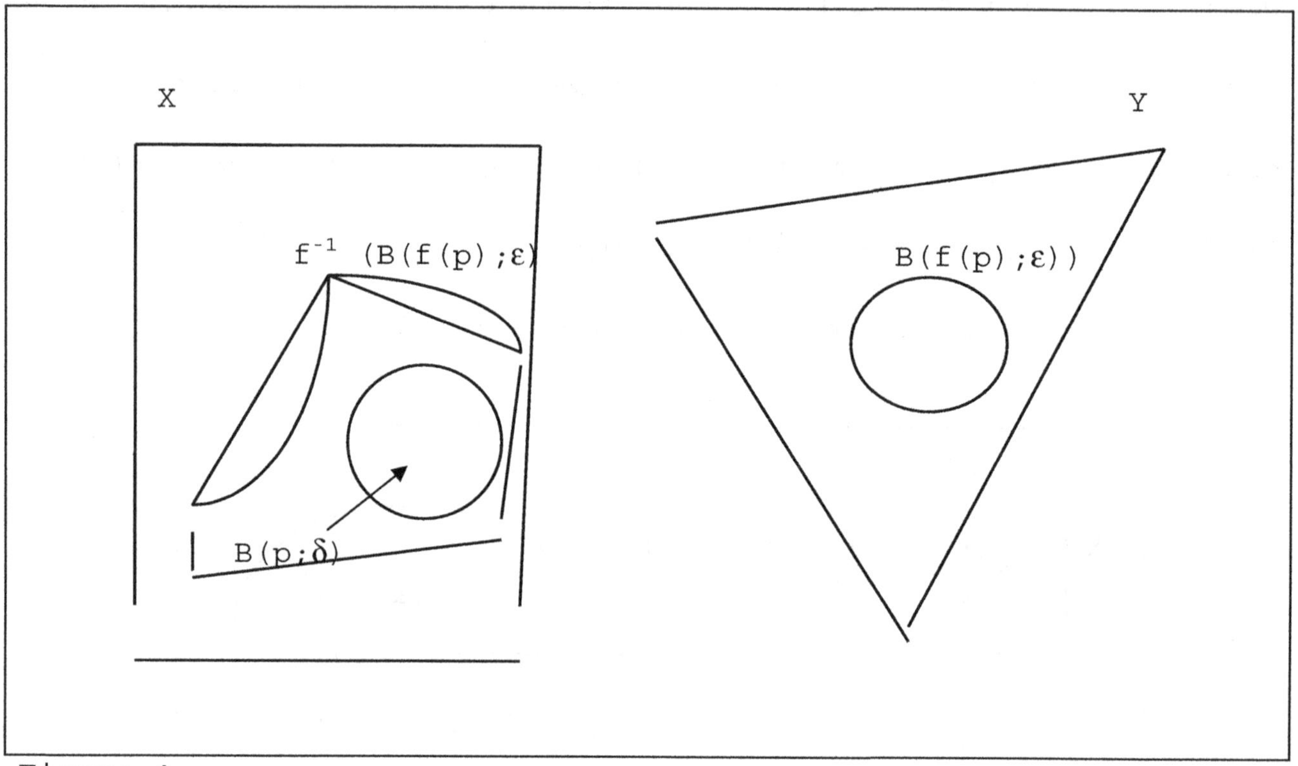

Figure 4

Neighborhood

2.4.4 Definition. Let (X,d) be a metric space and $P \in X$. A set $N \subset X$ is called a neighborhood (written nbhd) of p if there exists a $\delta > 0$ such that $B(p; \delta) \subset N$ (see Figure 5 below).

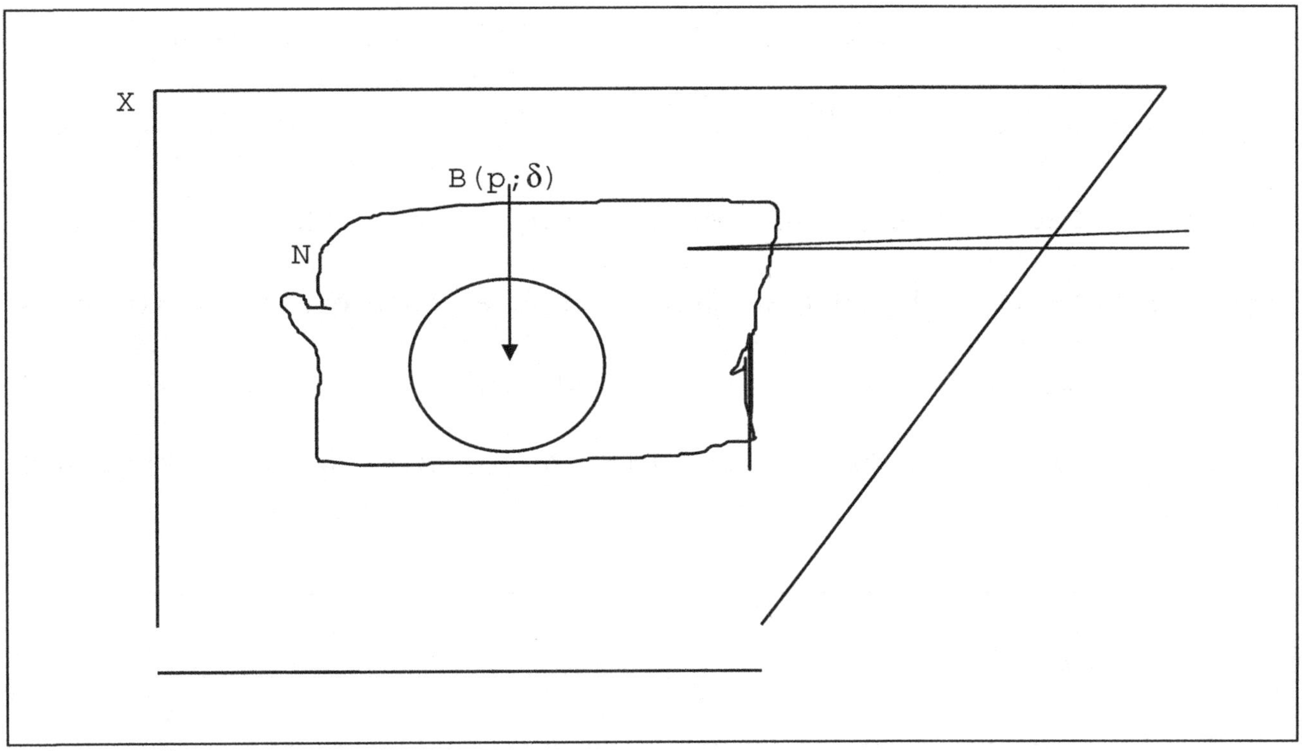

Figure 5

Remarks 2.4.2 The collection N_p of all nbhd's of a point p $\in$ X is called a complete neighborhood system of the point p. A neighborhood of a point p $\in$ X can be thought of as containing all of the points of X that are close to p or that it contains some open ball about p. In fact, for each $\in > 0$, B(p; $\in$) is a neighborhood of p. These open balls have the property that for each x $\in$ B(p; $\in$), B(p; $\in$) is a neighborhood of x.

2.4.5 Theorem: Let (X,d) be a metric space and p $\in$ X. For each $\delta > 0$, for each q $\in$ B(p; δ) there is a B(q; δ') so that B(q; δ') $\subset$ B(p; δ), that is the ball $B(p;\ \delta)$ is a nbhd of each of its points.

Proof. Let $q \in B(p;\ \delta)$, then $d(p,q) < \delta$. Choose $\delta' < \delta - d(p,q)$; note that $\delta - d(p,q) > 0$. If $x \in B(q;\ \delta')$ then $d(p,x) \leq d(p,q) + d(q,x) < d(p,q) + \delta' < d(p,q) + \delta' - d(p,q) = \delta$, therefore, $x \in B(p;\ \delta)$. Thus, $B(q;\ \delta') \subset B(p;\ \delta)$ (See Figure 6 below).

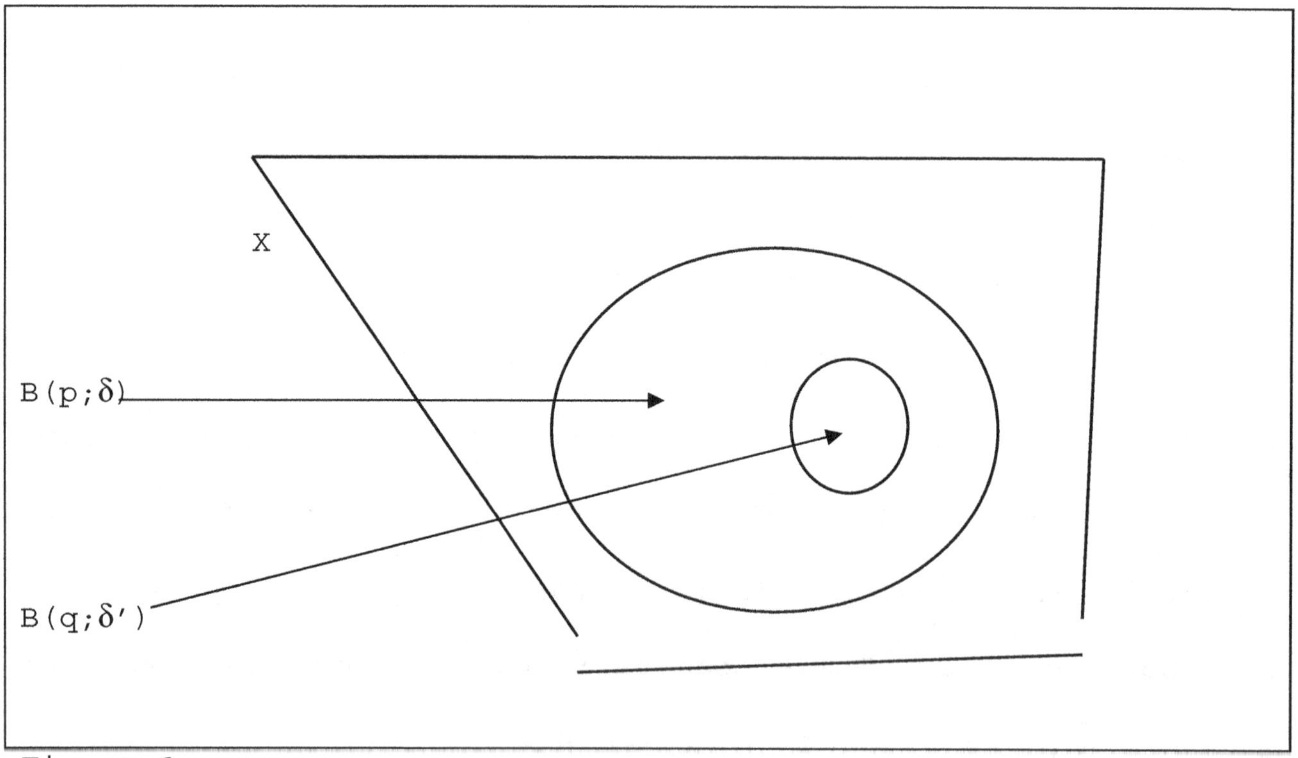

Figure 6

2.4.6 Theorem: Let (X,d) and (Y,d') be metric spaces, then $f:X\to Y$ is continuous at a point $p \in X$ if and only if for each neighborhood N of $f(p)$

there is a neighborhood N′ of p such that f(n′) ⊂ N.

Proof. (Exercise for the reader). We draw only helpful figures). See Figures 7 and 8 below.

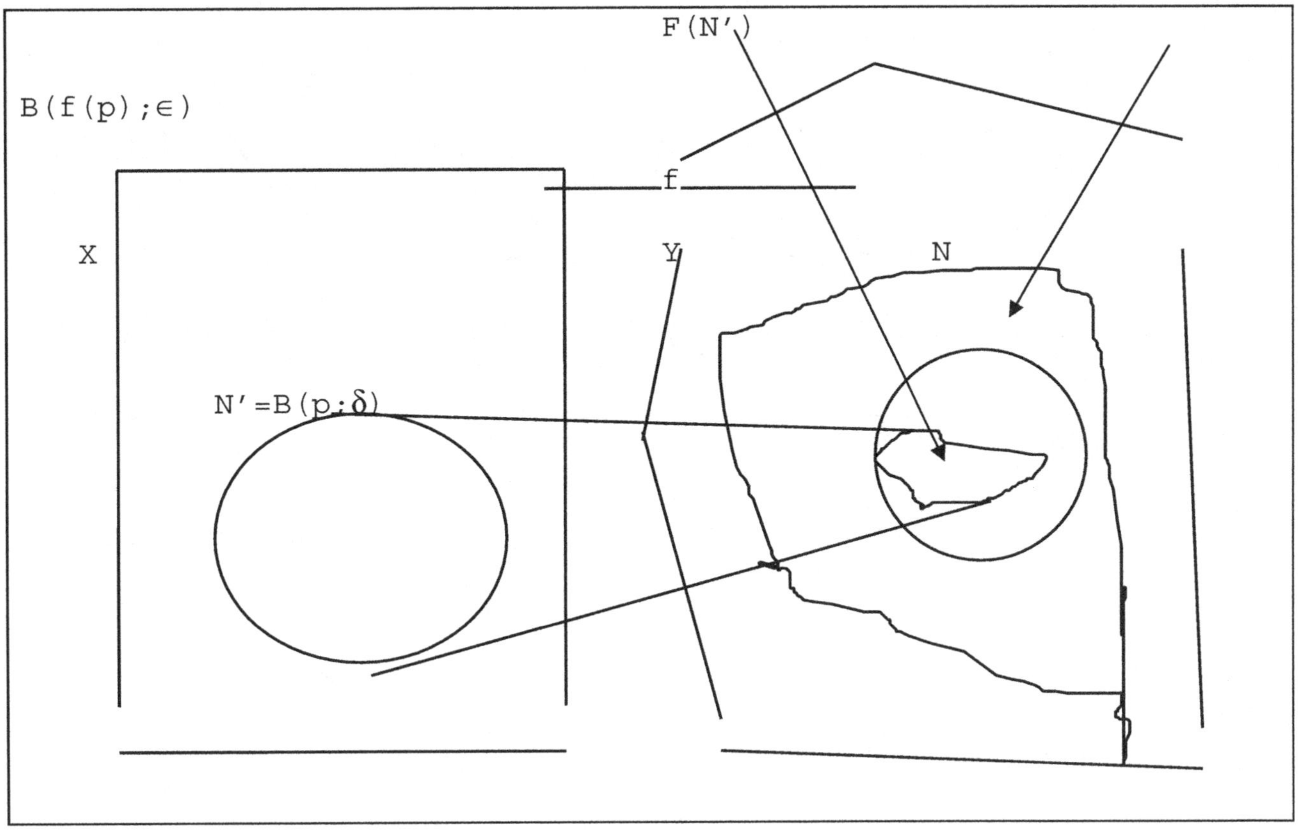

Figure 7 (For Proof ⇒)

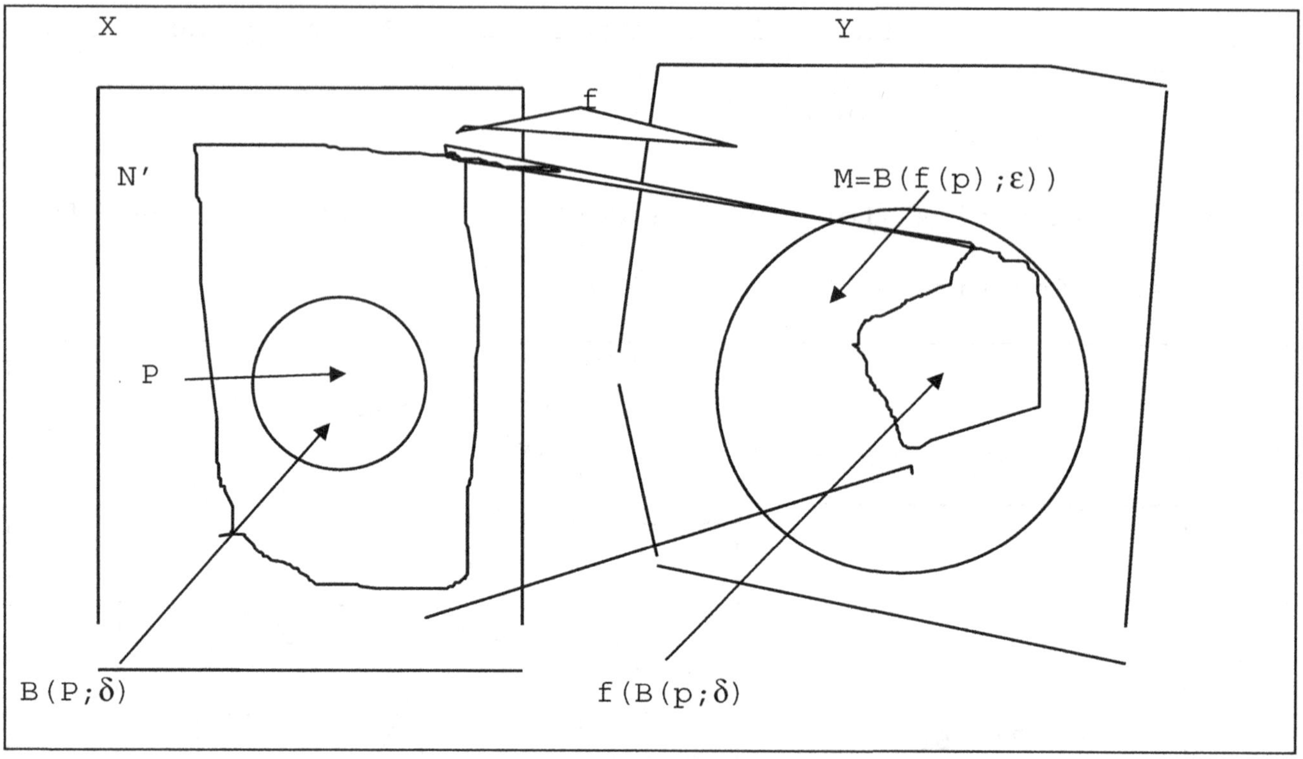

Figure 8 (For Proof ⇐)

2.4.7 Corollary. Let $f:(X,d)\rightarrow(Y,d')$ be a map. Then f is continuous at a point $p \in X$ if and only if for each neighborhood N of f(p) there is a neighborhood N′ of p such that $N' \subset f^{-1}(N)$.

Proof. (Exercise for the reader).

2.4.8 Theorem: Let (X,d) be a metric space; then we have the following:

(a) For each $p \in C$, there is at least one neighborhood of p.

(b) If N is a neighborhood of $p \in X$ and $N \subset M$, then M is a neighborhood of p.

Proof. (Exercise for the reader).

Basis

2.4.9 Definition. A collection N_p of neighborhoods of a point p of a metric space X is a basis for the neighborhood system at p if and only if every neighborhood N of p contains a member of N_p.

2.4.10 Theorem: Let (X,d) be a metric space. If M and N are neighborhoods of $p \in X$, then $M \cap N$ is a neighborhood if p.

Proof. (Exercise for the Reader). See Figure 9 below.

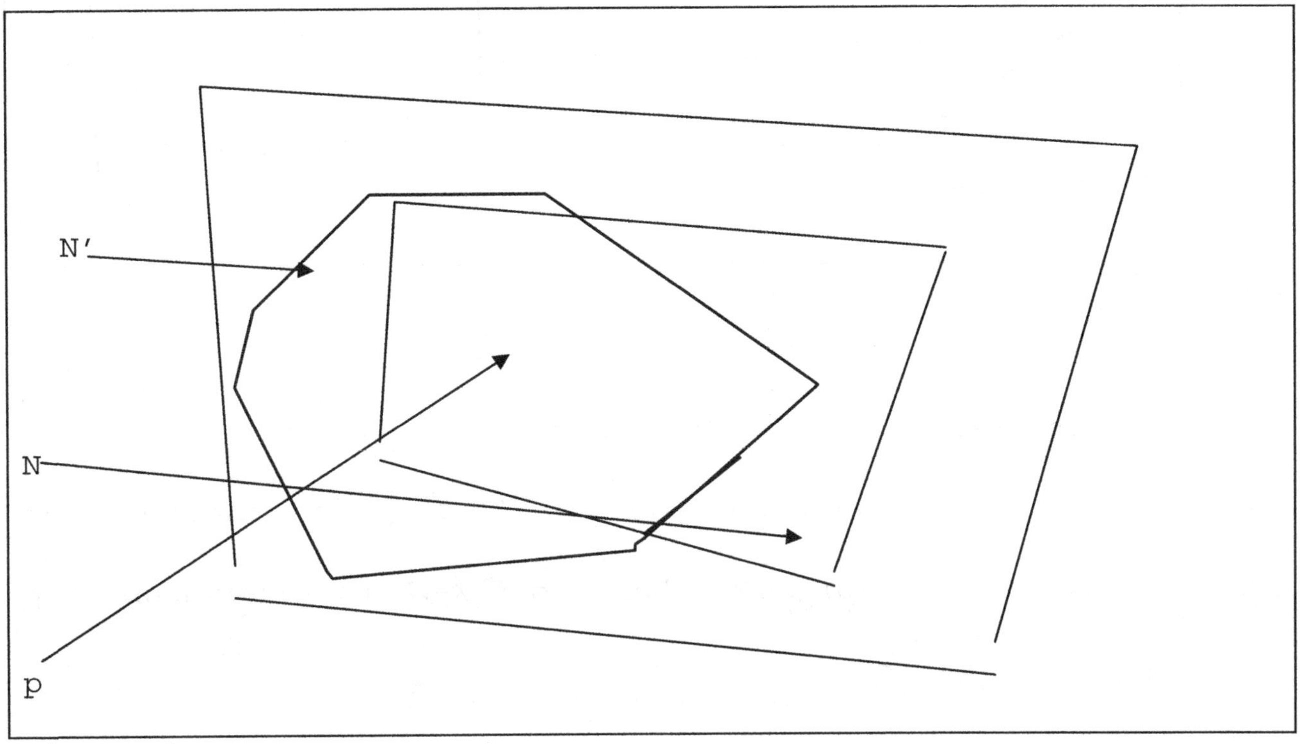

Figure 9

2.4.11 Lemma. Let $f:(X,d)\rightarrow(Y,d)$, then f is continuous at p $\in$ X if and only if for each neighborhood N of f(p), $f^{-1}(N)$ is a neighborhood of p.

Proof. (Exercise for the reader). See Figure 10 below.

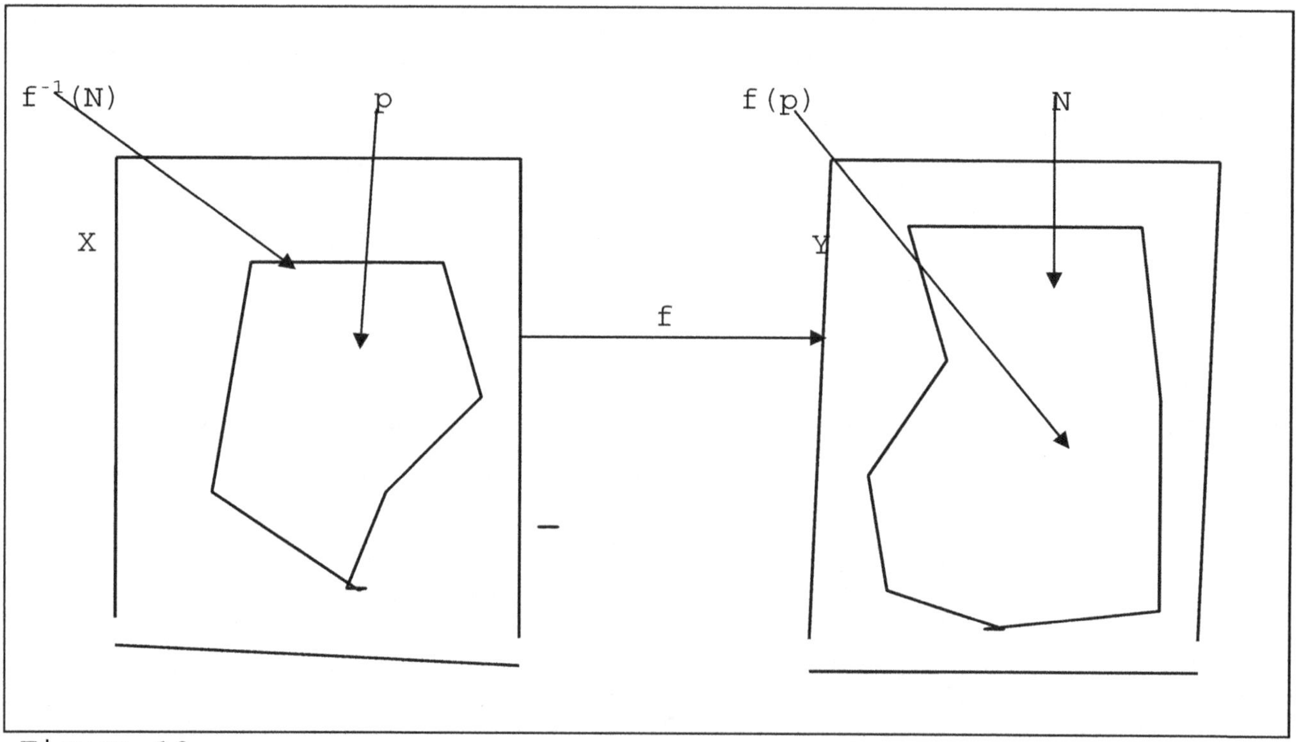

Figure 10

2.4.12 Theorem: Let $f:(X,d)\rightarrow(Y,d')$ be continuous at p $\in$ X and also $q:(Y,d')\rightarrow(Z,d'')$ be continuous at f(p) $\in$ Y, then g o $f:X\rightarrow Z$ is continuous at p $\in$ X.

Proof. (Exercise for the reader; Hint: use Lemma 2.4.11). See Figure 11 below.

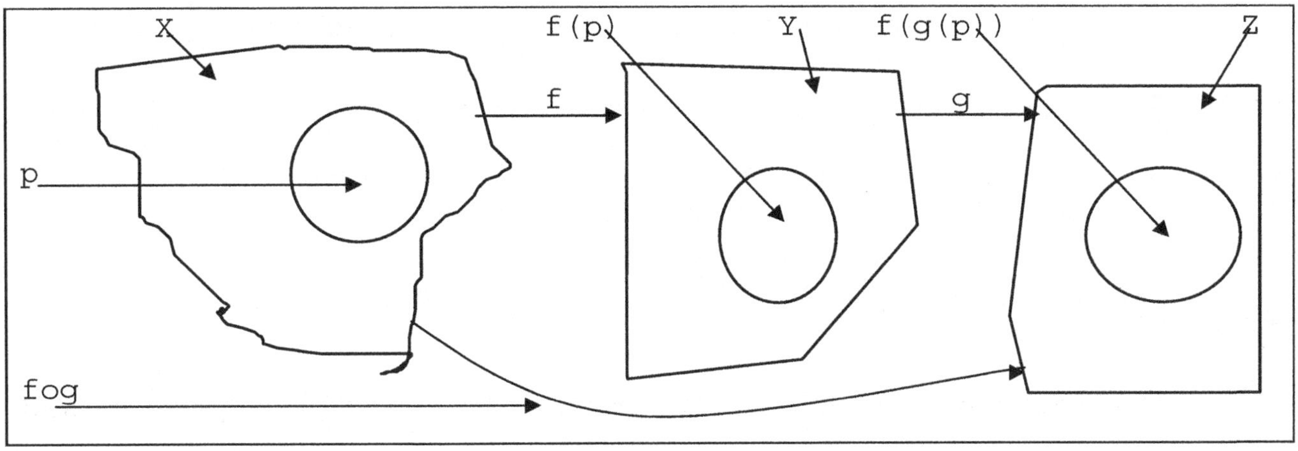

Figure 11

Exercises Set 9-4.

1. Let (X, d) be a metric space. Prove that for each $p \in X$ and each nbhd N of p, there exists a nbhd M of p such that $M \subset N$ and M is a nbhd of each of its points.

2. Let $p \in R$. Prove that $M \subset R$ is a nbhd of p if and only if M contains an open interval containing p.

3. Let $f:R \to R$ be continuous at $p \in R$ and $f(p) > 0$; then there is a number $d > 0$ such that $f(x) > f(p)/3$ when $|x - p| < d$.

4. Let $F: [0, 1] \to [0, 1)$ be a continuous and onto. Is it true that $f(x) = x$ for some $x \in [0, 1]$?

5. Is onto needed in exercise 4? Prove your claim?

6. Is continuity needed in exercise 4? Prove your claim.

2.5.0. Open, Closed Sets in a Metric Space.

In a metric space it is clear that $B(p; \delta)$ is a nbhd of each of its points. We called $B(p, \delta)$ an open ball about p of radius

δ. It is clear that we want open sets to be characterized by , because the collection of subsets possessing this property (nbhd of each of its points) plays an important role in topology.

Open Set

2.5.1 Definition. A subset U of a metric space is open if and only if U is a nbhd of each of its points.

2.5.2 Theorem: Let (X, d) be a metric space; then:

(a) $U \subset X$ is open if and only if U is the union of the open balls.

(b) X and $\varnothing$ are open in (X, d).

Proof. (a) ($\Rightarrow$) Suppose U is open, then for each $x \in U$, there is a $B(x;\ \epsilon_x) \subset U$ and $B(X;\ \epsilon_x)$ an open ball about x. Thus $U = \cup_{x\in U} B(x;\ \epsilon_x)$.

(a) ($\Leftarrow$) If $U = \cup_{x\in I} B(x;\ \epsilon_x)$, $u \subset X$. If $x \in U$, then $x \in B(x;\ \epsilon_x)$ for some $x \in I$. $B(x;\ \epsilon_x)$ is a nbhd of x since $B(x{:}\ \epsilon_x) \subset U$, U is a nbhd of x. Thus, U is a nbhd of each of its points and by definition U is open.

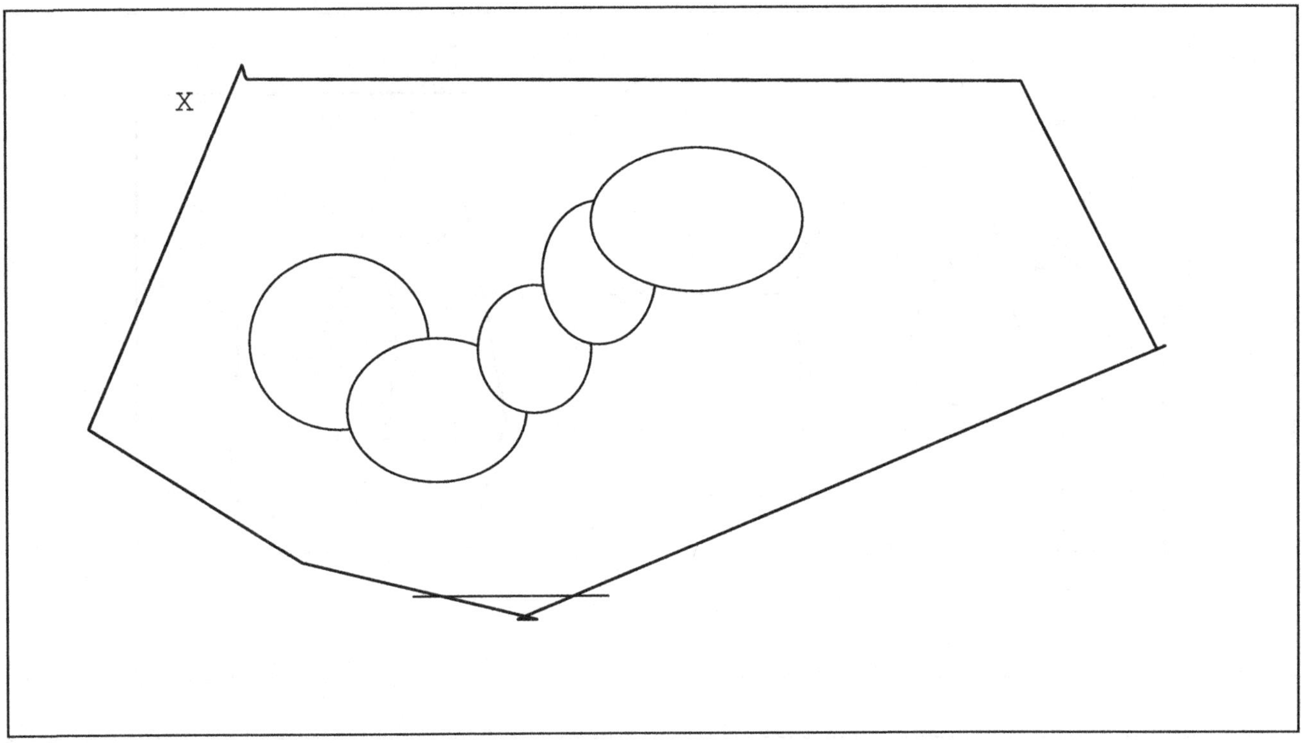

Figure 12

(b) (Exercise for the reader)

2.5.3 Theorem: Let $f:(X,d)\to(Y,d')$; then f is continuous if and only if for each open set U of Y, $f^{-1}(U)$ is open in X.

Proof. (Exercise for the reader; we draw figures only). Hint: take $p \in f^{-1}(U)$ and show that $f^{-1}(U)$ is a neighborhood of p.

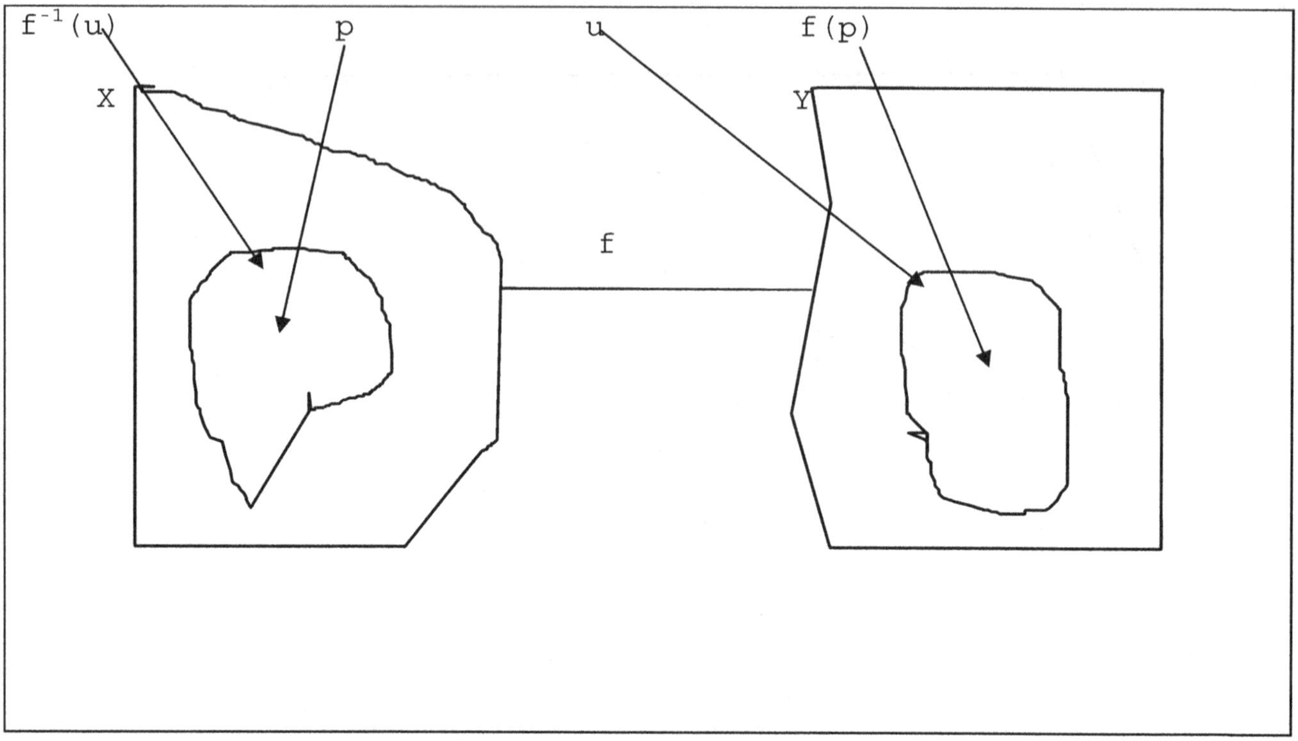

Figure 13 (For Proof $\Rightarrow$)

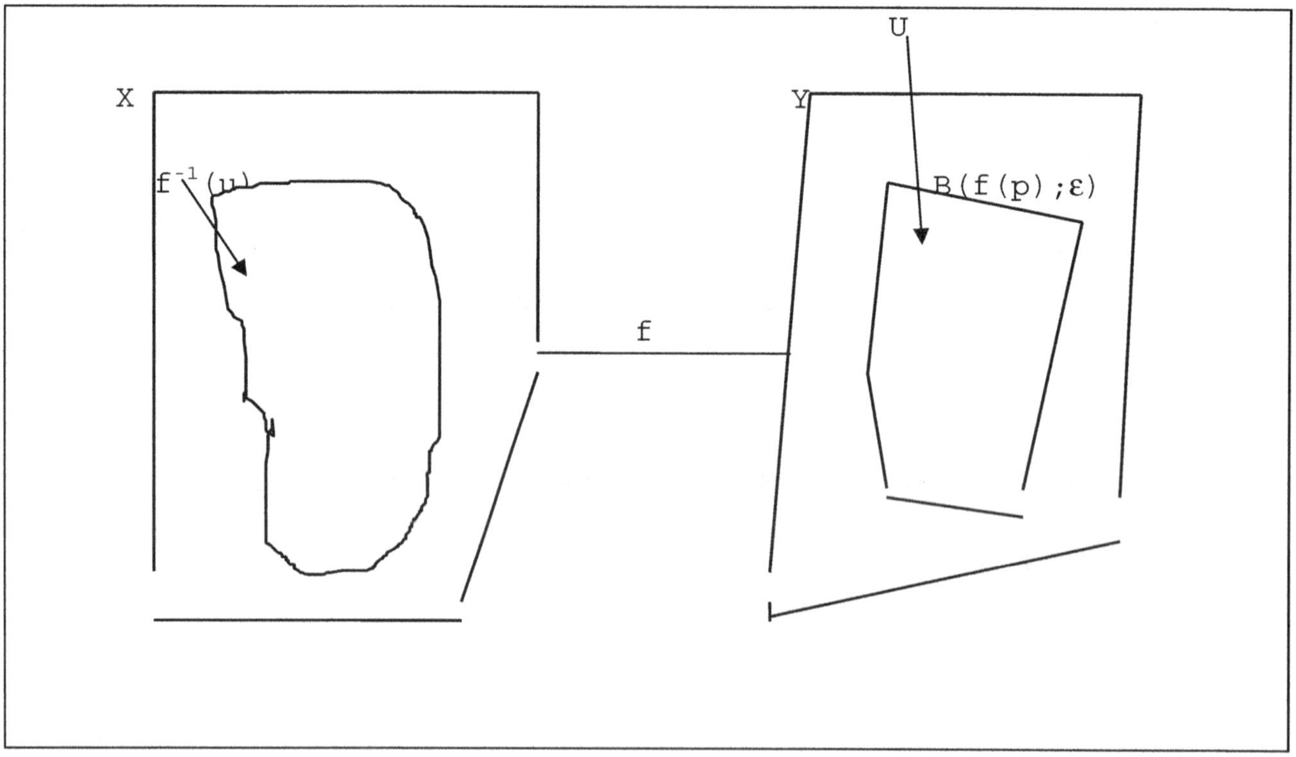

Figure 14 (for Proof $\Leftarrow$)

Remark 2.5.1 We have considered a point $p \in X$, where (X, d) is a metric space and all points with in $\in$ of p as an arbitrary degree of closeness. The concept "limit point of a sequence" is important. First let us define this concept as applied to the real numbers.

<u>Limit Point</u>

2.5.4 Definition. Let p_1, p_2, ..., p_n be a sequence of real numbers. A real number p is the limit point of this sequence if for each $\in > 0$, there is a $N > 0$, N an integer such that, $|p_n - p| < \in$ whenever $n > N$. We write $\lim_{n\to\infty} p_n \to p$ and say that p_n converge to p.

Remark 2.5.2 Now let us consider the metric space (X, d) and a sequence p_1, p_2, ..., p_n of X. In this case we would like to deal with the degree of closeness of p_1, and p, P_2 and p,..., p_n and p where $p \in X$. We now formulate our definition for metric spaces.

<u>Limit For Metric Space</u>

2.5.5 Definition. Let (X, d) be a metric space and P_1, P_2, ... be a sequence of points in X. A point $p \in X$ is said to be the limit point of this sequence if $\lim_{n\to\infty} d(P_n, p) = 0$ or $\lim_{n\to\infty} p_n = p$.

Example 1. Let $X = R$ and $d(a, b) = |a - b|$ then (X, d) is a metric space. Let $p_n = 1/n$ and n be a positive integer then $\lim_{n\to\infty} d(P_n, 0) = 0 \therefore \lim_{n\to\infty} p_n = 0$.

2.5.6 Theorem: Let (X, d) be a metric space and P_1, P_2, ..., be a sequence of points in X. Then $\lim_{n\to\infty} p_n = p$ for $p \in X$ if and only if for $\in > 0$, there is an integer N, such that $p_n \in B(p, \in)$ whenever $n > N$.

Proof. (Exercise for the reader; Hint: Use definition.)

2.5.7 Corollary. Let (X, d) be a metric space and p_n be a sequence of points in X. Then

(a) $\lim_{n\to\infty} p_n = p$ for some $p \in X$ if and only if for each $n_2$hd u of p there is an integer N such that $p_n \in U$ when $n > N$.

(b) $\lim_{n\to\infty} p_n = p$ for some $p \in X$ if and only if for each set U with $p \in U$ there is an integer N such that $p_n \in n > N$.

Proof: (Exercise for the reader; Hint: Use Theorem 2.5.6).

2.5.8 Theorem: Let (X, d) be a metric space and p_n a sequence in X with $\lim_n p_n = p$ and $\lim_n p_n = q$ and $p, q \in X$, then $p = q$.

Proof. Suppose $p \neq q$ then $d(p, q) = \in > 0$ and let $U = B(p; \in/2)$, $V = B(q; \in/2)$, (see Figure 16 below). Then U and V are nbhds of p and q respectively and $U \cap V = \emptyset$; thus by Theorem 2.5.6 we have for some N $P_n \in U$, $n > N$ and $P_n \notin V$ for $n > N$. Therefore $\lim_n p_n \neq q$ contradiction hence $p = q$.

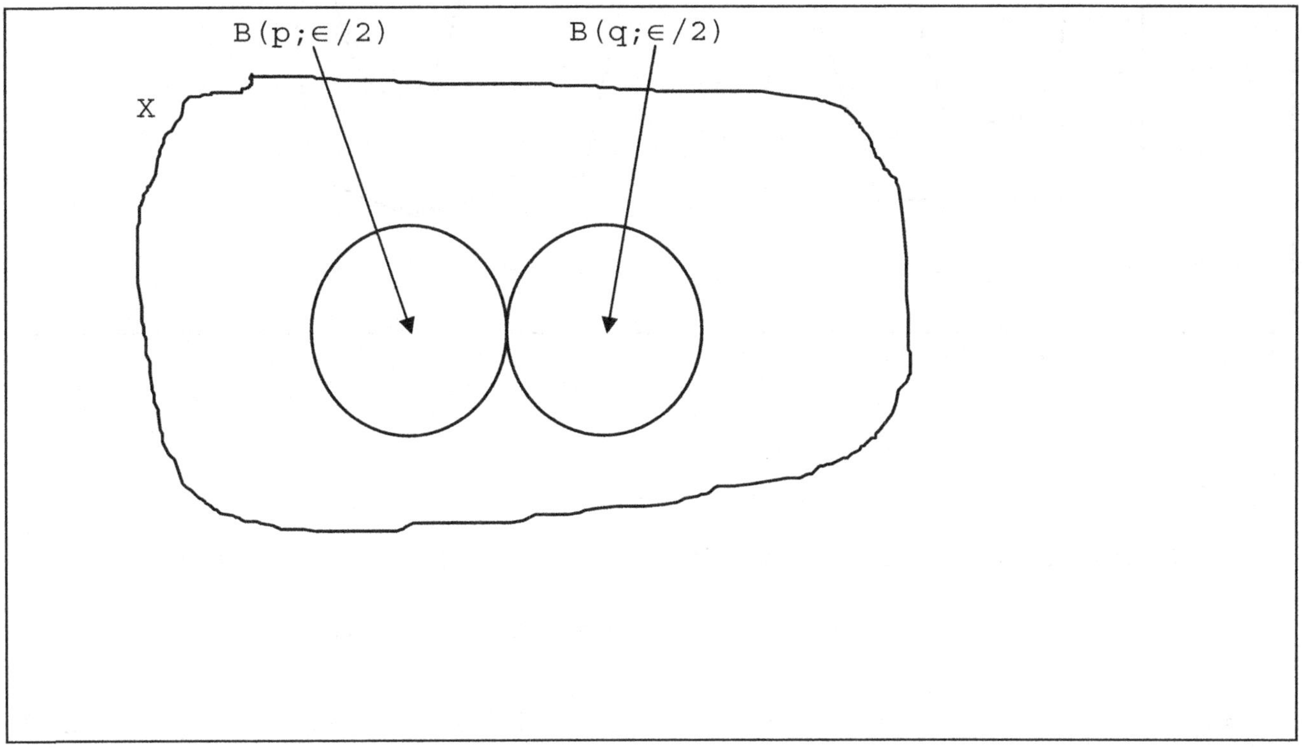

Figure 15

2.5.9 Theorem: Let $f:(X, d) \rightarrow (Y, d)$ then f is continuous at $p \in X$; if and only if, whenever $\lim_n p_n = p$ for a sequence p_n of X, $\lim_n f(p_n) = f(p)$.

Proof. (Exercise, we draw figure only).

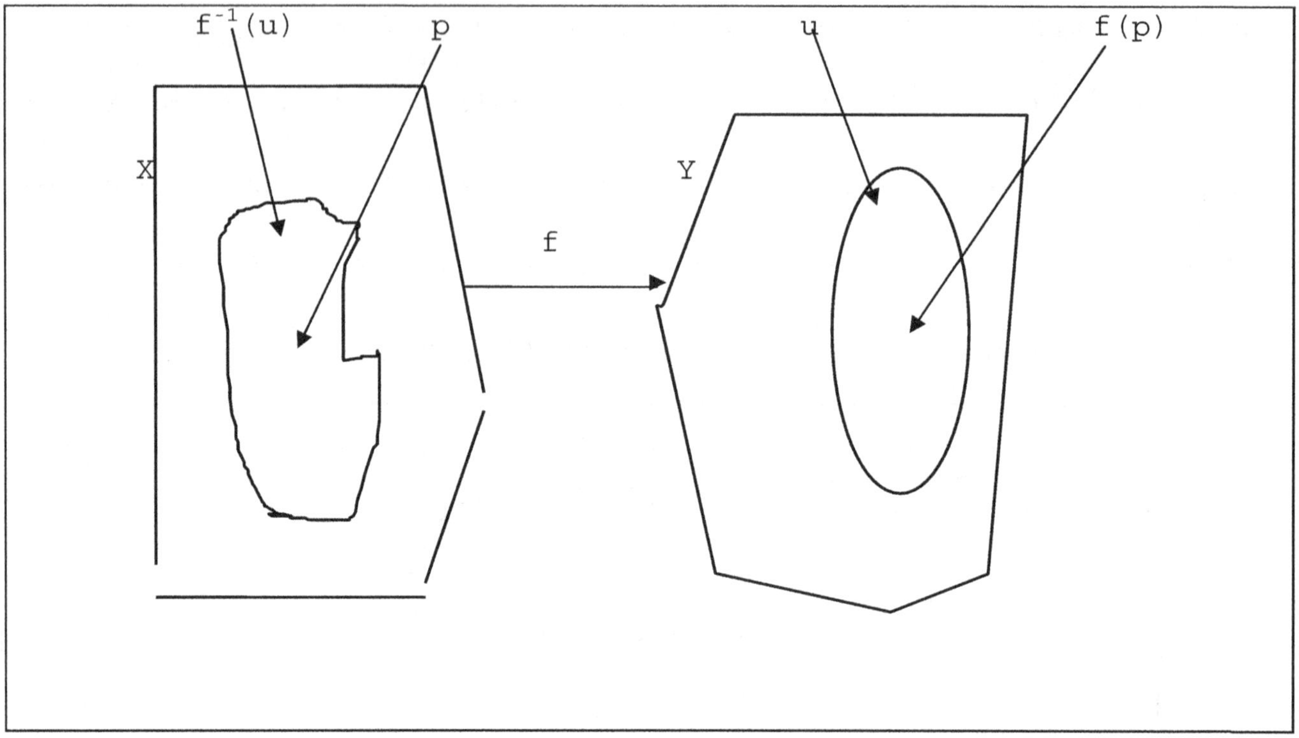

Figure 16 (for Proof ⇒)

Closed Set

2.5.10 Definition. Let (X,d) be a metric space. A subset S is closed if and only if X\S is open.

Examples 2.5.1 ∅ is open in X and hence X\∅ = X is closed in X. Also, ∅ is closed in X, since X is open in X and ∅ = X \ X. Thus, we do have sets that are both open and closed. We will see other examples of sets that are both open and closed. Also, since (a, ∞) and (-∞, -a) are open in R we have that U = (-∞ - a) U (a, ∞) is open in R ∴ [-a,a] = X\U is closed in R.

Remark 2.5.3. It is clear from Definition 5.9.1 that a closed set contains all of its limit points. Also, an open set contains no limit point of its compliment. We can also make a connection between continuity and closed sets.

2.5.11 Theorem: Let $f:(X,d)\rightarrow(Y, d')$; then f is continuous if and only if for each closed set $S \subset Y$, $f^{-1}(S)$ is closed in X.

Proof: (Exercise for reader; Hint: Use the fact that $f^{-1}(Y\backslash S = X\backslash f^{-1}(S))$.

Remark: 2.5.3 Let (X, d) be a metric space. Let $p \in X$ and S be a non-empty subset of X. We have $d(p,S) = glb\{b(a,x): x \in S\}$ (See Exercise 2.3.1).

Example 2.5.2 Let X = R, S = n (4,3) then d(10,S) = d(10,3) = $|10-8| = 2$

<u>2.6.0 Subspace of A Metric Space</u>

If (X, d) is a metric space, here we wish to consider the best way to convert $Y \subset X$ into a metric space. It would appear that the best way to do this would be to restrict $d: X \times X\rightarrow R$ tp $d:Y \times Y\rightarrow R$.

<u>Subspace</u>

2.6.1 Definition. Let (X,d) and (Y,d) be metric spaces; we say that (Y,d') is a subspace of (X,d) if:

(a) $Y \subset X$

(b) $d' = d|y \times Y$.

Example 4 Let Q be the set of rational numbers. Define $d' = d|Q \times Q$ where $d(x,y) = |x - y|$ for each $x,y \in R$ then (Q, d') is a subspace of (R,d).

Remark 2.6.1 From the example above it is clear that there are as many subspaces of a metric space (X,d) as there are nonempty subsets of X.

2.6.2 Theorem: Let (Y,d') be a subspace of (X,d) then

(a) For each $p \in Y$, $B_Y(p; \delta) = Y \cap B_X(p;$ is $\delta)$.

(b) For each $p \in Y$, $N' \subset Y$ is a neighborhood of p if and only if there exists a neighborhood N of p in (X,d) such that $N' = Y \cap N$.

Proof. (Exercise for the reader).

2.6.3 Theorem: Let (X,d') be a subspace of (Y,d). A subset $U' \subset X$ is open in (X,d') if and only if there exists an open subset $U \subset Y$ such that $U' = X \cap U$.

Proof. (⇒) Suppose U′ is open in (X,d′); then for each $p \in U'$, there is a $\delta_p > 0$ such that $B_x(p;\ \delta_p) \subset U'$. The set $B_y(p;\ \delta_p)$ is open in (Y,d) and thus the set $U = \cap_{p \in U'} B_y(p;\ \delta_p)$ is open in (Y,d). But note that

$$
\begin{aligned}
X \cap U &= X \cap (\cup_{p \in U'} B_y(p;\ \delta_p)) \\
&= \cup_{p \in U'} x \cap B_y(p;\ \delta_p)) \\
&= \cup_{p \in U'} B_x(p;\ \delta_p) \\
&= U'
\end{aligned}
$$

(See Figure 18 below).

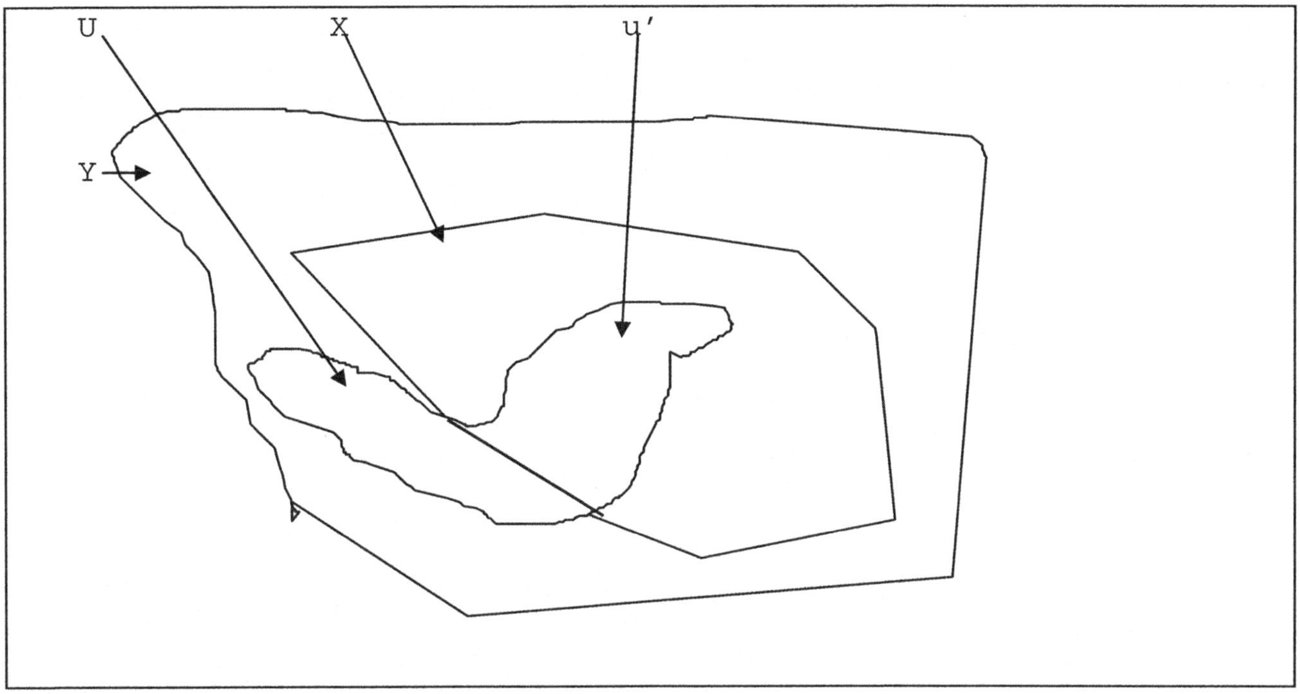

Figure 17

(⇒) Suppose U is open in (y,d) and $U' = X \cap U$. For each $p \in U'$, since $p \in U$, there is a $\delta_p < 0$ such that $B_y(p;\ \delta_p) \subset U$.

Then $p \in B_X(p;\ \delta_p) = X \cap B_Y(p;\ \delta_p) = U'$. Why? Hence, U′ is open in (X,d′).

Chapter 3

Topologies, Topological Spaces and Mapping on Topological Spaces

3.1.0 Introduction

In this chapter we will define topological spaces in general, what is meant by topology on a set, and look at some examples of topological spaces, as well as mapping on topological spaces and some topological ideas.

Topology is the study of intrinsic qualitative properties or certain kinds of sets which hereafter we shall call spaces. An intrinsic qualitative property roughly means one that is unaltered by stretching and/or bending without tearing. For example, let $X = [1, 2]$ and $f:x \to R$ be defined by $f(x) = 2x-1$; then $f(x) = [1,3] = y$, and one may say that f is a mapping that stretches X into Y (See Figure 1 below).

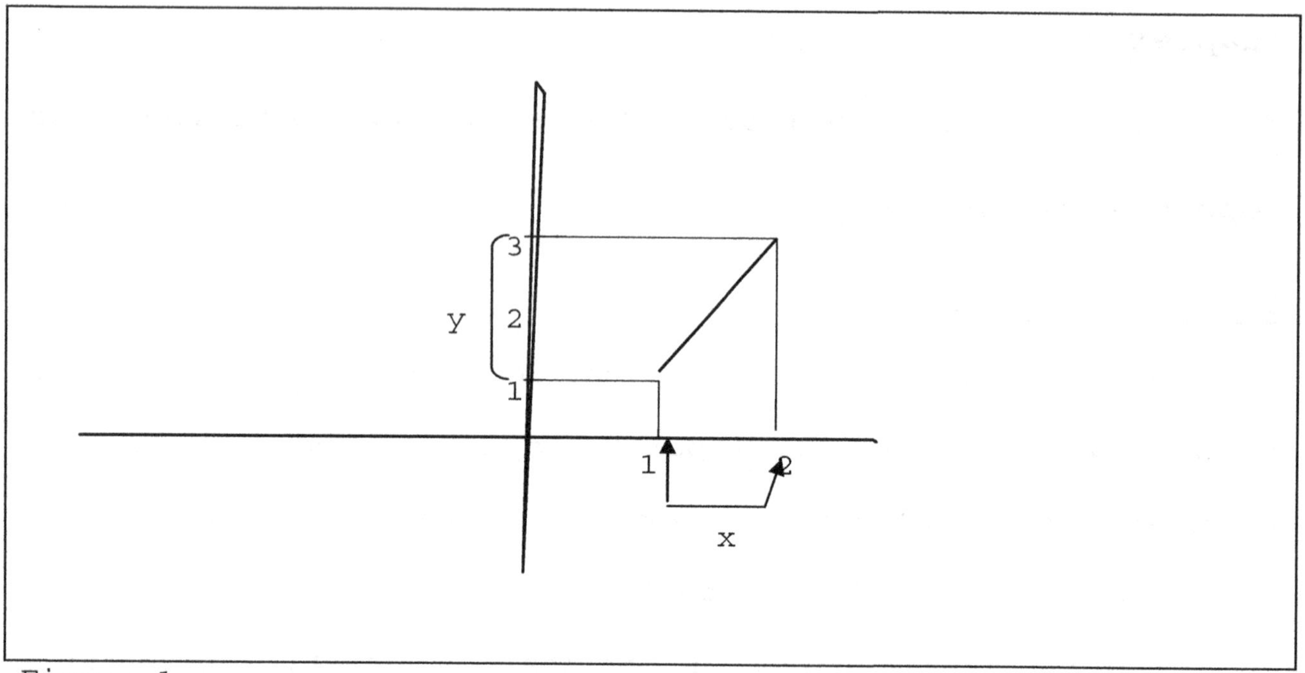

Figure 1

One should note that stretching or compressing need not be linear. For let x - [o, π/2] and let g: X→R defined by g (x) = cos x then g(X) = [o, 1] = y, and one may say that g is a mapping that compresses X into y (See Figure 2 below).

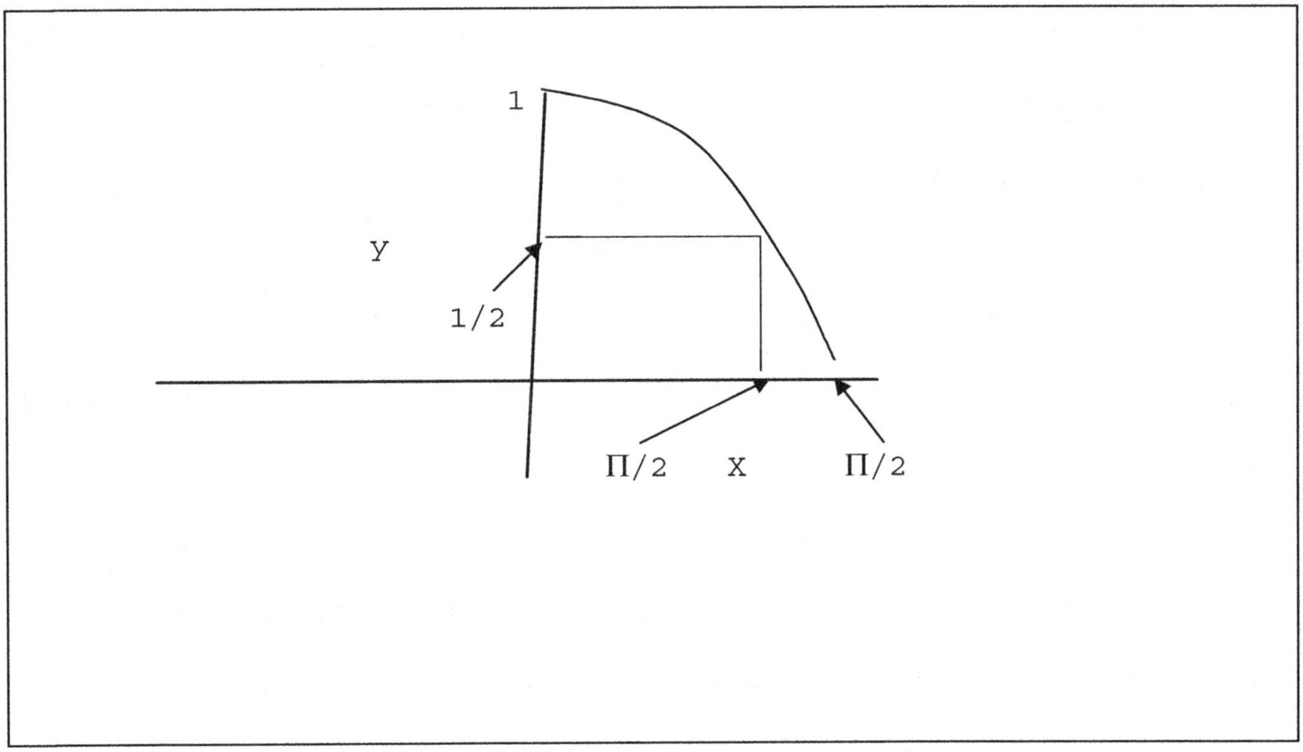

Figure 2

To keep things in order, the reader should keep familiar spaces (especially the real line and metric spaces) in mind when he/she studies the definitions and proves theorems, and the reader should draw pictures of the situation at hand; however, keep in mind that a picture is not a proof.

Discarding the distance function and retaining the open sets of a metric space will give rise to a new mathematical concept, called a *topological space*. Here we will introduce the topological concepts that we studied in Chapter 2 in the context or topological spaces. When we drop the distance function and retain the idea of open sets, we will find that, in most cases what is a Theorem in a metric space will become a definition in a topological space.

There are other ways to introduce topological spaces. If, upon dropping the distance function of a metric space, we retain a system or neighborhood of points of the metric space, we obtain what we shall call a neighborhood space.

The idea of continuity is very important. In Chapter 2, the author extended the definition of a function to a mapping from one metric space to another. A topology is the minimum structure on a set that will allow us to talk about continuity.

In this chapter, we abstract the essential properties of open sets to obtain a new concept, a topology on a set.

The reader should keep in mind familiar spaces (especially the real line and more general metric spaces) when he/she studies the definitions and proves the theorems, and should draw pictures representing and illustrating the situation at hand (with the understanding that a picture is not a proof).

Many of the concepts introduced and theorems proved in metric spaces can be reformulated without regard to metric but solely in terms of open sets. these carry over easily to the new, more general notion of topological space which we define below. However, for some notions such as continuity which are defined alternative characterizations. These theorems can serve as the basis for definitions in the more general context.

3.2.0 Definition and Examples of Topological Spaces

In Theorem 2.5.2(b) of Chapter 2 we proved that both X and Y are open in a metric space (X, d), and that the intersection of any finite collection of open subsets of (X,d) is open. Also, the union of any family of open subsets of (x,d) is open. We abstract these properties and make the following definition.

Topological Space

3.2.1 Definition. Let X be a non-empty set. A topology on x is a collection of subsets T of X that satisfies the following three conditions:

1. $\varnothing$, X $\in$ T
2. If $A_1, A_2, A_3, \ldots, A_n \in T$, then ${}^{n}\cap_{i=1} A_i \in T$
3. If $A_i \in T, \forall \in I$, then $\cup_{i\in I} A_i \in T$

   The pair (X, d) is called a topological space, and the subsets of X that are members of T are called open sets in (X,T). If it does not lead to confusion, we will just say open subsets of X and say space X other than space (X, T).

Remark 3.1.1 When we are dealing with a topological space, in some cases the particular topology that a space has may not be important, and results that we obtain in this case are valid for any topological space. Hereafter we will write space X and not space (X, T). In some cases circumstances may require that we write space (X, T). Note that a set $u \in X$ is open in X means that $u \in T$. The following is a strategy for drawing a figure, similar to a Venn diagram to illustrate a finite topology on a set.

Strategy For Illustrating A Finite Topology On A Set A By The Use of A Figure

1. Draw a rectangular region to represent A. If A is finite a dot may be used for each member of A (see example).
2. Draw a loop inside A to represent each member of the topology other than A.
3. The empty set has no representation (see example).

The resulting figure is always an illustration for the given topology on A (see the following illustration).

Example 3.2.1

Let $A = \{a, b, c\}$ and $T = \{\emptyset, A\}$. Show that (A, T) is a topological space and draw a figure to illustrate the topology.

Solution:

1. Note that $\emptyset$ and A are in T.
2. What are the finite subcollections of T? They are $\emptyset$, $\{\emptyset\}$, $\{A\}$, and $\{\emptyset, A\}$. The only one we need to check

is {∅, A} since T only has two members, and $\emptyset \cap A = \emptyset \in T$.

3. What are the families of subcollections of T? They are only the ones listed in 2 above and $\emptyset \cap A = \emptyset \in T$.

Since we have satisfied Definition 3.2.1, we have that T is a topology on A and (A, T) is a topological space. If we use the strategy as outlined above we get Figure 3.

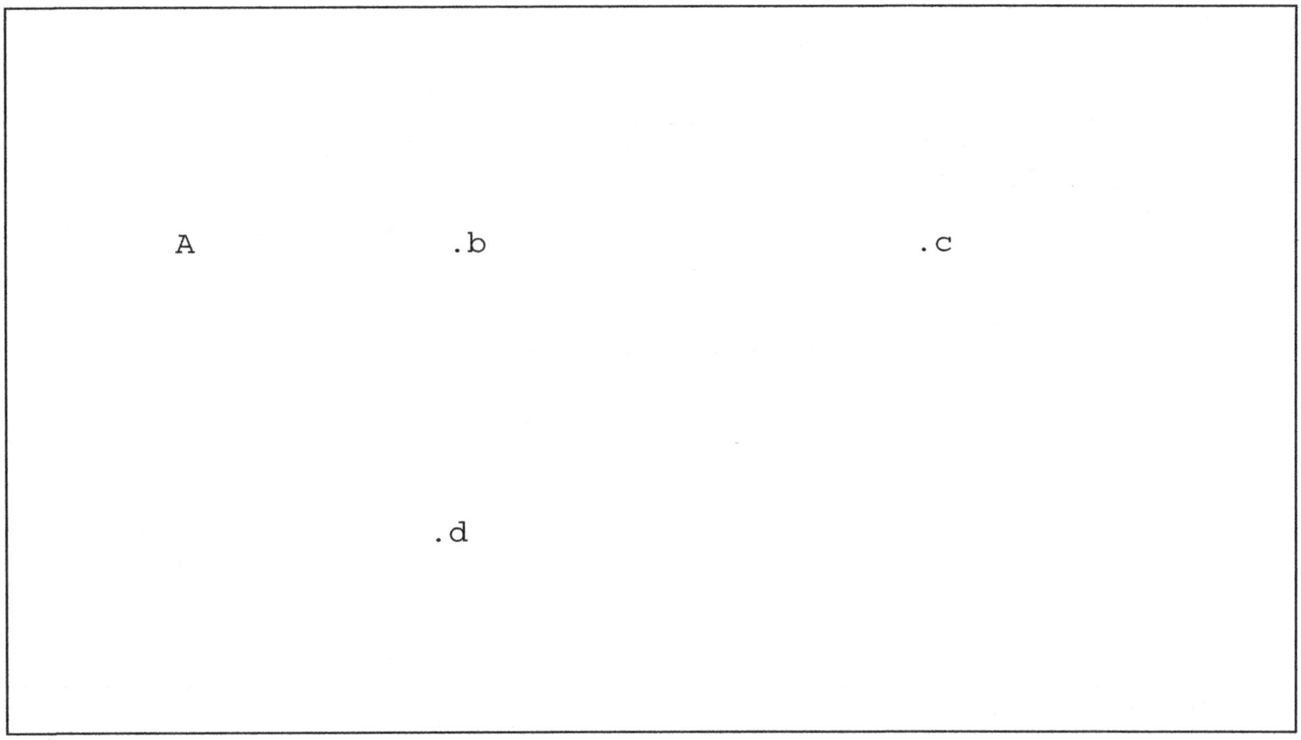

Figure 3

Note that Figure 1 can be explained as follows:

1. The topology has only two members ∅ and A.

2. The empty set has no sketched region. The sketch for A is the whole space or region. Thus the figure is correct.

Example 3.2.2

Let $A = \{b, c, d\}$ and $T = \{\emptyset, A, \{b, c\}\}$. Show that $\{A, T\}$ is a topological space.

Proof. Exercise for the reader; Hint: we draw Figure 4).

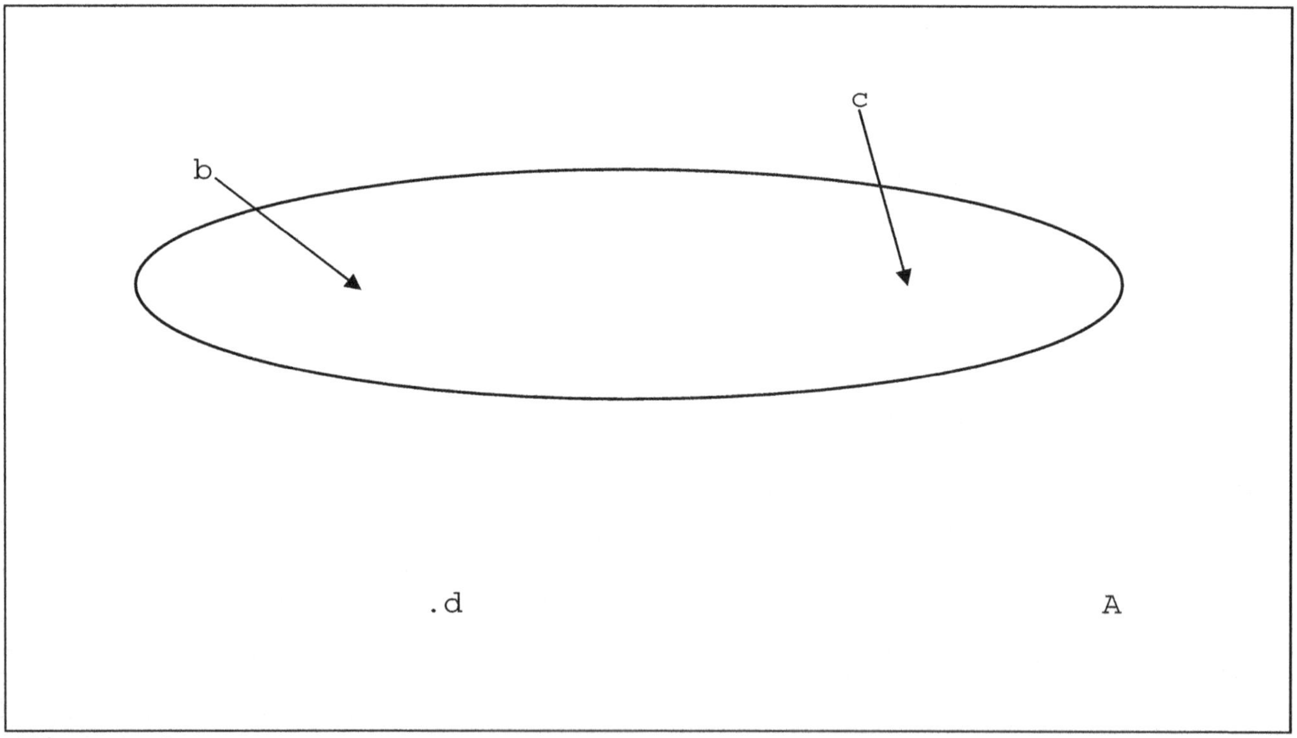

Figure 4

Note that Figure 2 can be explained as follows:

1. Since A is the only set in T containing d, d is included only inside the region that represents A (in this case a rectangle).

2. Since b and c are contained in two subsets of T, namely A and {b, c}, there is a region inside the region representing A that includes b and c.
3. The empty set is represented by no region and A by the whole rectangular region.

Example 3.2.3

Let (A, d) be any metric space and T = {U ⊂ A: U open in (A, d)}. Show that (A, T) is a topological space.

<u>Solution</u>:

Since we have shown that ∅ and a are open, the intersection of finitely many open sets in a metric space is open and the union of any collection of open sets in a metric space is open, we have that (A, T) is a topological space.

Non-Example 3.2.1.

Let A = {b, c, d} and T = {∅, {b, c,},{c, d}, A}. show that (A, T) is not a topological space. Draw a figure.

<u>Solution</u>:

1. ∅ and A are in T.
2. What are the finite subcollections in T? They are: ∅, {∅}, {{b, c}}, {{c, d}}, {A},{∅, {{b, c}}, {∅, {c, d}}, {∅, A}, {{b, c}, {c, d}}, {{b, c}, A}, {{c, d}, A}, {∅, {b, c}, {c, d}}, {∅,

{b, c}, A}, {{b, c}, {{c, d}, A}, {∅, {b, c}, {c, d}, A}, {∅, {c, d}, A}.

Note that:

$\varnothing \cap \{b, c\} = \varnothing \in T$

$\varnothing \cap \{c, d\} = \varnothing \in T$

$\varnothing \cap A = \varnothing \in T.$

however $\{b, c\} \cap \{c, d\} = \{c\} \notin T$. Hence, we have an intersection of a finite subcollection of T that is not in T. Thus T is not a topology.

What does it mean to say that T is not a topology for A? *At least one* of the conditions in 2.1 is violated.

Condition 1 is violated it *either* $\varnothing \notin T$ *or* $A \notin T$, i.e., if either one (or both) of ∅, A is not in T.

Condition 2 is violated if there is *at least one* finite subcollection of T whose intersection is not in T. In the non-example above {{b, c}}, {{c, d}} was such a collection

Condition 3 is violated it there is *at least one* subcollection of T whose *union* is not in T. In non-example 1 {{b, c}}, {{c, d}} was such a subcollection.

All it takes for a set not to be a topology is for one of the conditions to be violated!

We illustrate T by Figure 5.

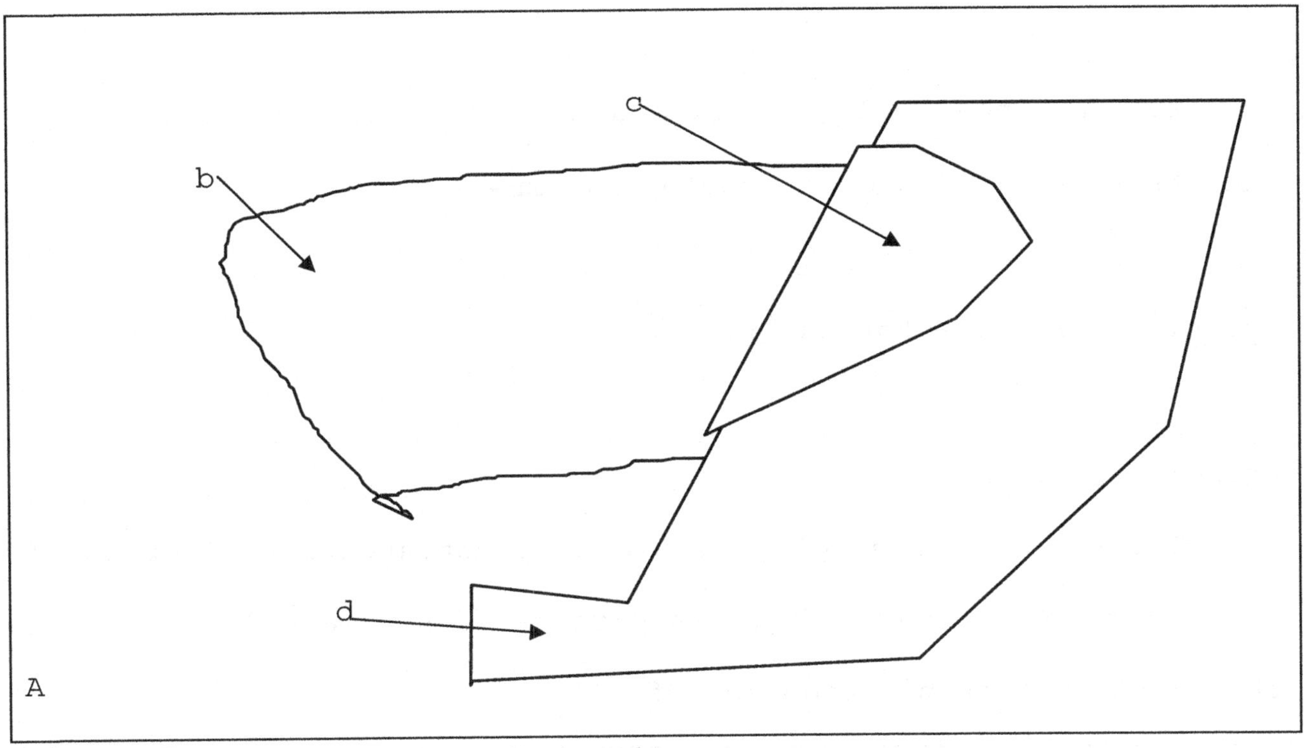

Figure 5

The rectangle represents A, the oval enclosing b and c represents {b, c} and the oval enclosing c and d represents {c, d}, $c \in \{b, c\} \cap \{c, d\}$ as is evident from the figure but there is no loop around c indicating that {c} is not in T. Hence, T is not a topology.

Problem 3.2.1

Let A = {b, c}, find all topologies on A.

<u>Solution</u>:

Topology 1 = {∅, A}

Topology 2 = {∅, {b}, A}

Topology 3 = {∅, {c}, A}

Topology 4 = {∅, {b},{c}, A}

Non-Example 3.2.2

Let A = the real numbers and T = {U ⊂ A: A\U is finite}. show that (A, T) is not a topological space.

Solution: Exercise for the reader.

Exercise 3.2.1

Let A = {b, c, d, e}. Figure 4 illustrates a collection of subsets of A. Which if any additional sets must be added so that it illustrates a topology of A?

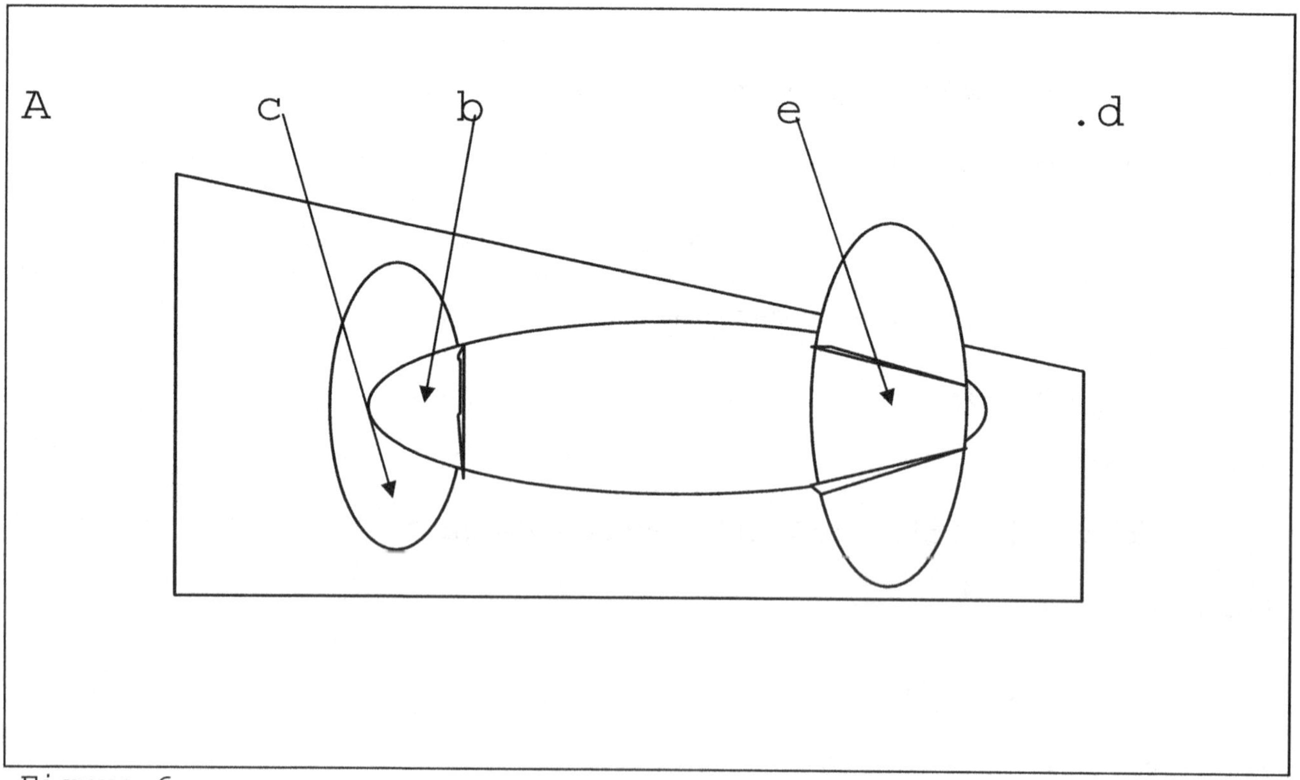

Figure 6

Solution:

Exercise for the reader.

Non-Example 3.2.3

Let $A = \{b, c, d, e\}$. Let $T = \{\varnothing, A, \{b, c,\}, \{c, d\}\}$ then (A, T) is not a topological space. What must be added to T to make it a topology on A? Draw a figure to illustrate your explanation.

Solution:

Exercise for the reader.

In each example show that T is a topology in X. Draw figures when possible.

Example 3.9.1 Let $X = \{0,1,2\}$ then take

$T = \{\varnothing, X, \{o\}$ then $(X; T\}$

is a topological space.

Example 3.2.5 Let X be any non-empty set, and let $T = \{\varnothing, X\}$, (X, T) is called the indiscrete topology on X.

Example 3.2.6 Let X be any set and let $d = P(X)$ (the power set of X); then (X, T) is called the discrete topology on X. Recall that $P(X) = \{u: u \subset X\}$

3.2.2 Theorem: Let X = R, R the real numbers, and let T = {U: R\u is finite or} then (X, T) is a topological space.

Proof: Exercise for the reader.

Remarks 3.2.2. As we have seen in Chapter 2, we can define many metrics on a set. We also can define many topologies on a set X. Check the following examples to make sure that they are topologies on X.

Example 3.2.7 Let X = {o, 1}, T = {{o}, ∅, X}, then (X, T) is a topological space (see Figure 7 below).

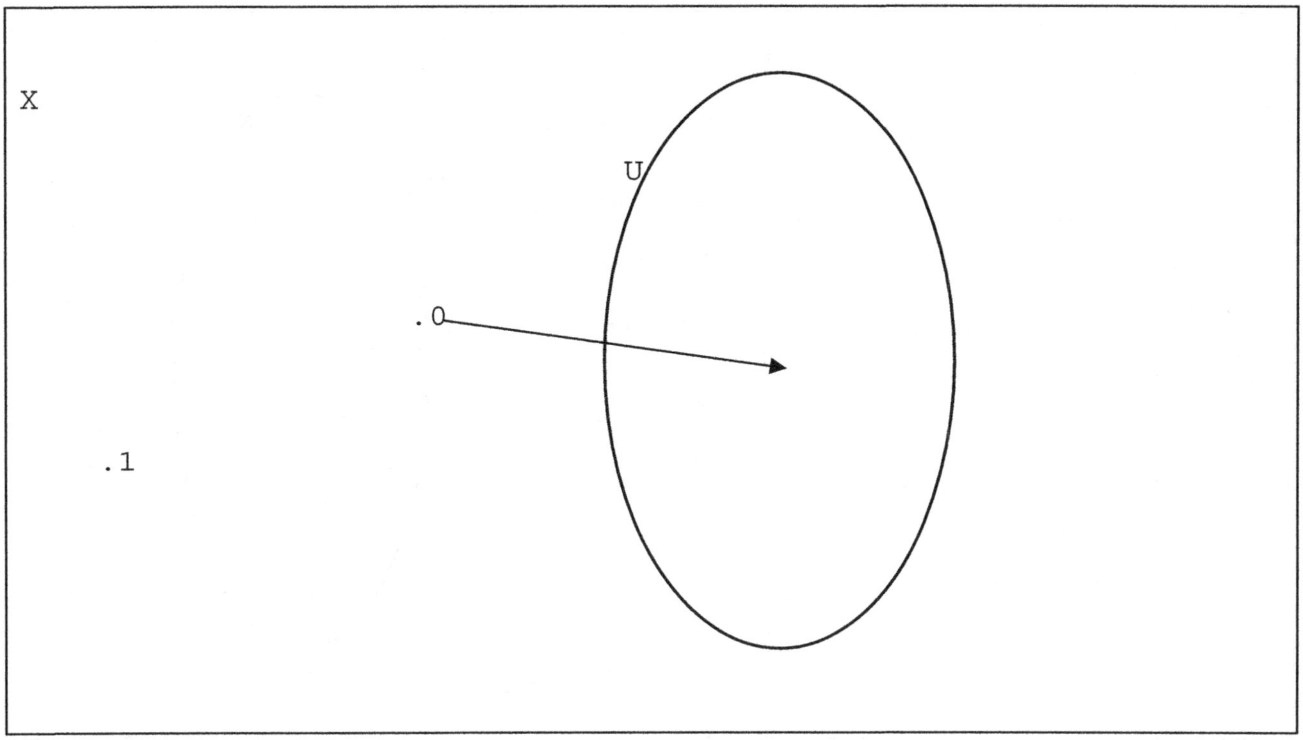

Figure 7

Example 3.2.8 Let {a, b}, T = {∅, {a} X}; then (X, T) is a topological space (see Figure 8 below).

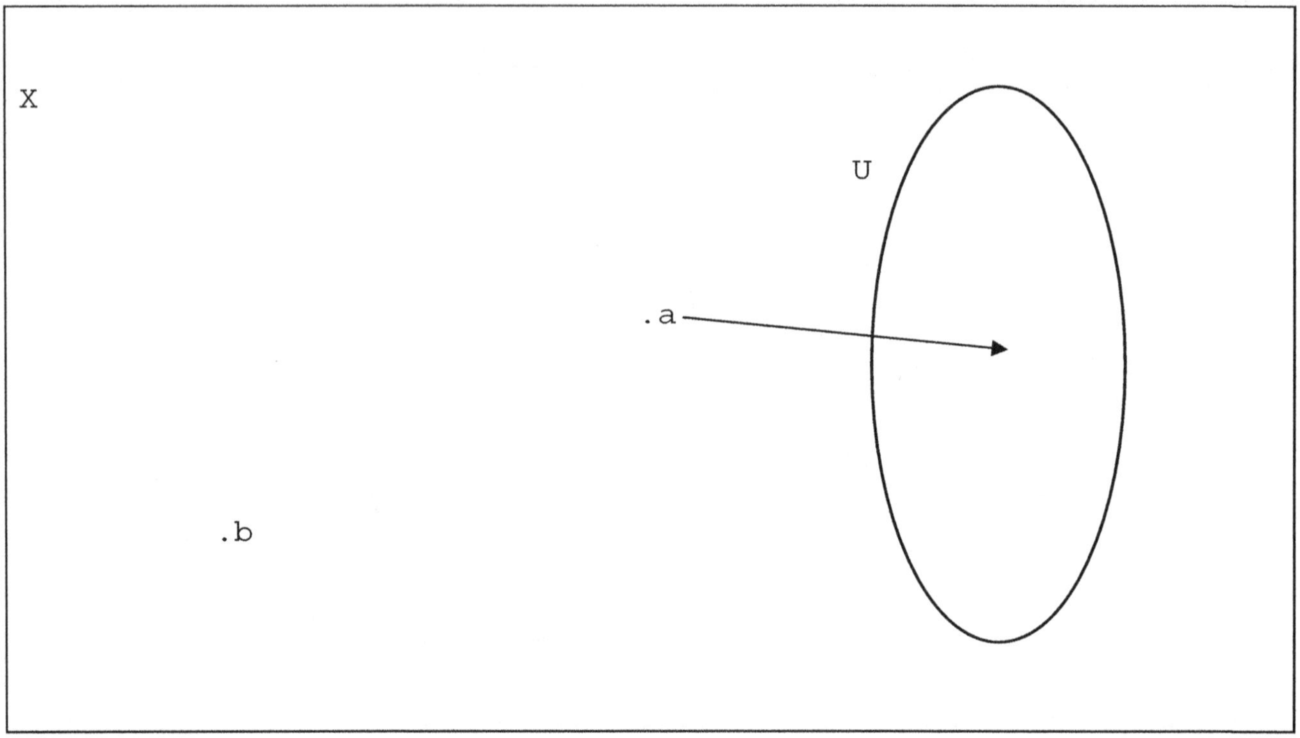

Figure 8

Example 3.2.9 Let X = {a, b}, T = {∅,{a},{b},X}; then (X, T) is a topological space (see Figure 9 below).

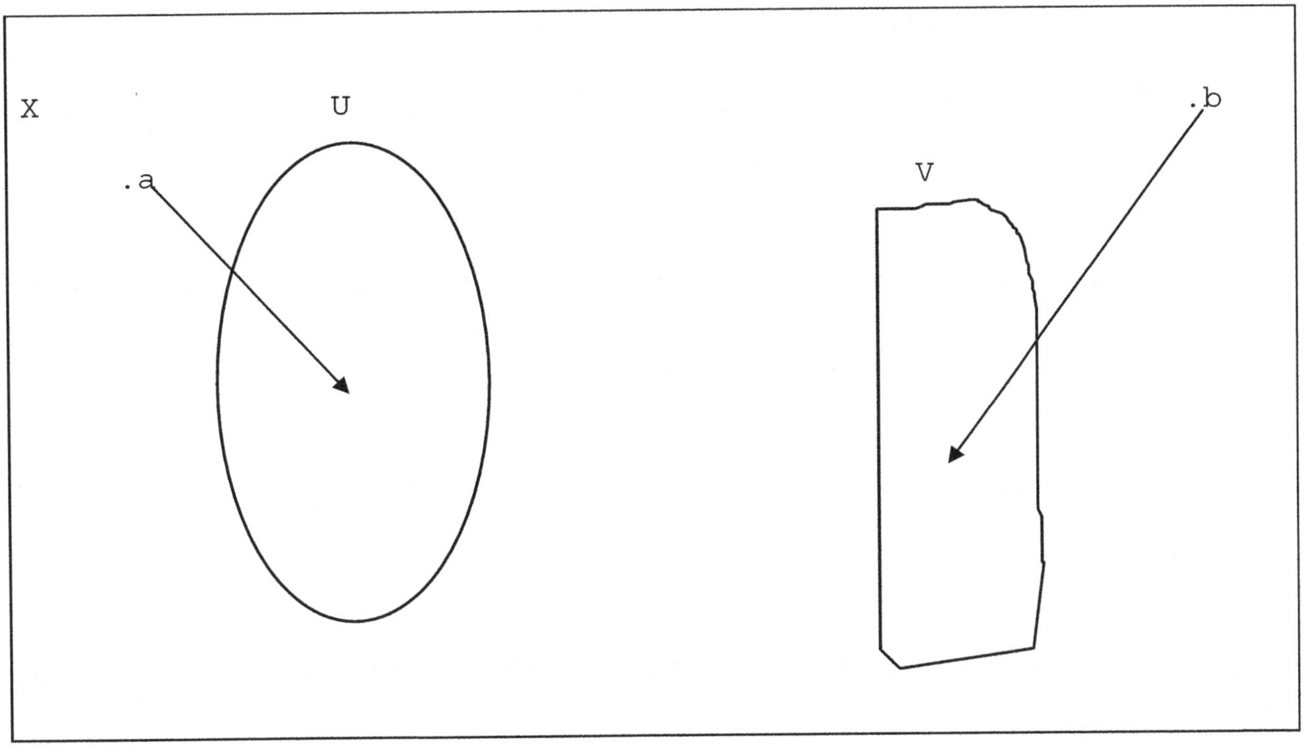

Figure 9

3.2.3 Theorem: Let R be the set of real numbers. Let T = {ϕ} ∪ (X \ A: A is a countable subset of R. Show that T is a topology for R.

Proof: Exercise for the reader.

<u>Exercises Set 3-2</u>

1. Let X = {0, 1, 2}. Give all the topologies on the set X.

2. Let X be a set. Show that the indiscrete topology, the discrete topology and this finite complement topology are in fact topologies on X.

3. Let X = {a, b, c} and let T = {∅, {a}, {b}, X}. Show that (X, T) is not a topological space.

4. Let X = {a, b, c} and let T = {∅, X, {a, b},{b, c}}. Show that d is not a topology for X.

5. Let X be any set and T = {U: X\U is either countable or all of X}. Show that T is a topology on X.

6. Let X be the set of real numbers. Prove that each of the following is a topology on X.

   (a) T′ = {U ⊂ X: if P ∈ U, then there is an interval [a, b) with P ∈ [a, b) ⊂ U}.

   (b) T′′ = {U ⊂ X: if P ∈ U, then there is an interval (a, b) with P ∈ (a, b) ⊂ U.

   (c) T′′′ = {U ⊂ X: if P ∈ U, then there is an interval (a, b], with P ∈ (a, b] ⊂ U.

Note: The topologies T′, T′′, and T′′′ in Exercise 6 are called the right-half open, usual, and left-half open topologies on the set X respectively.

7. Let X be the set of natural numbers, and let the members of T_x, be defined as follows:

$0_0 = \varnothing$

$0_1 = X$

$0_2 = \{2, 3, 4, \ldots\}$

$0_3 = \{3, 4, 5, \ldots\}$

.

.

.

$0_n = \{n, n+1, n+2 \ldots\}$

Prove that (X, T_x) is a topological space.

3.3.0 Metrizable Topological Spaces

Figure 8 indicates the relationship between metric spaces and topological spaces. It must be noted that two metric spaces (X,d) and (X, d') may give rise to the same topological space (X,δ). Also, there are topological spaces (X', δ'), which could not have arisen form any metric space.

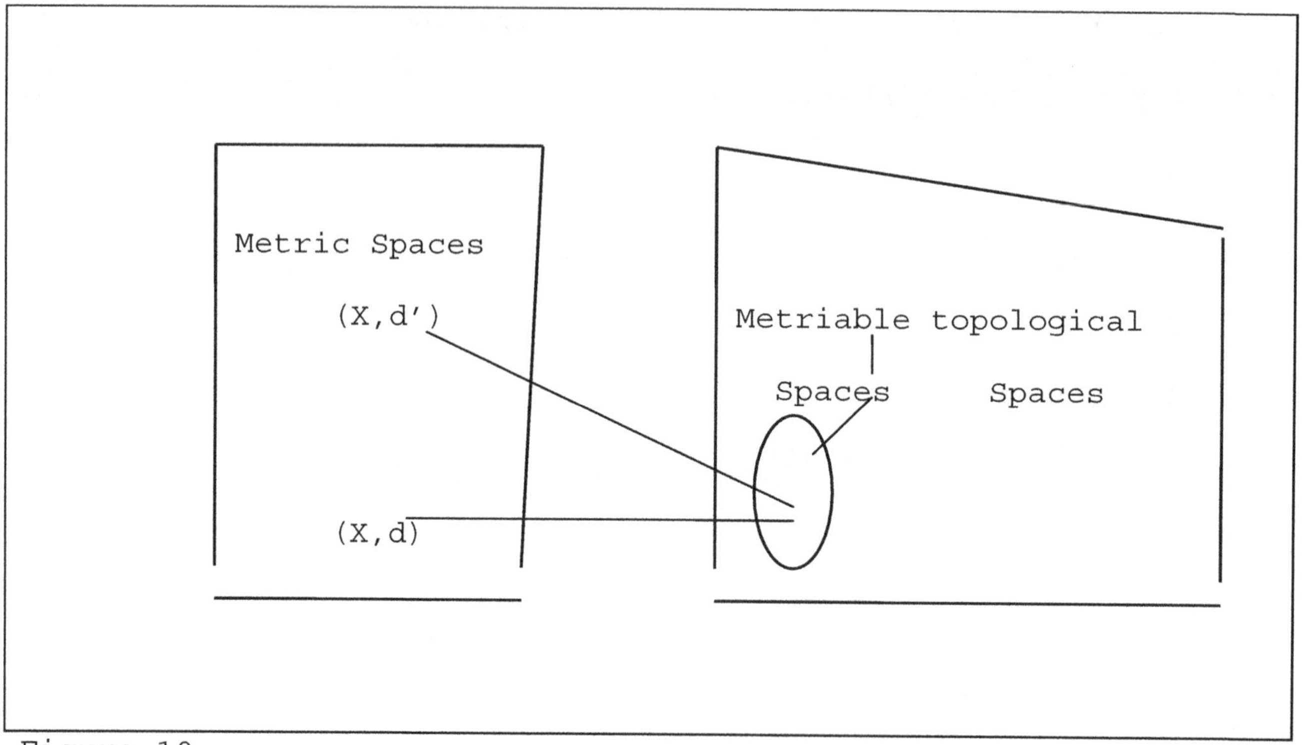

Figure 10

Metric Space

3.3.1 Definition: The collection of topological spaces that arise from metric spaces are called the collection of metrizable topological spaces, and the open sets in the metric space are also open in the topological space.

3.3.2 Theorem: Let (X, d) be a metric space; then T = (U: U is open in (X, d)} is a topology on X.

Proof. We need only satisfy Definition 2.2.1. $\emptyset$ is open, for if $\emptyset$ is not open, then there would have to be a $P \in \emptyset$.

If P $\in$ X, then for $\delta > 0$, $B(x; \delta) \subset X$ and, hence, X is open.

Let $U_1, U_2, \ldots U_3 \in T$) and take $p \in .\cap_{i=1}{}^{n} u_i$ Then since u_i is a nbhd of p for each i=1, 2,..., n, by Theorem 4.3 $\cap_i{}^n{}_{=1} U_i$ is open in (X, d), hence $\cap^n{}_{i=1} U_i \in T$

Finally, let $U = \cup_{i \in A} U_i$ and let $p \in U$ where for each $i \in A$, U_i is open. Then $p \in U_\propto$ for some $\propto \in A$ and $U_\propto$ is a nbhd of p. Since $U_d \subset U$, is a nbhd of p. therefore, U is a neighborhood of each of its points; hence, U is open in (X, d) and $U \in T$. Thus, T is a topology in X or (X, T) is a topological space.

Neighborhood In a Topological Space

3.3.3 Definition: Let (X, T) be a topological space; a subset N of X is a neighborhood of a point $p \in X$ if and only if there is a $U \in T$ with $p \in U \subset N$.

3.3.4 Theorem: Let (X, T) be a topological space. A subset U is open in X if and only if U is a neighborhood of each of its points.

Proof. (Exercise for the reader).

Closed Set

3.3.5 Definition: Let (X, T) be a topological space; a set $S \subset X$ is closed if an only if X\S is open in X.

Example 3.1.1 [1, 2] is closed in R, since [1, 2] = R\{(-∞ 1)∪(2, ∞)} and (-∞ 1)∪(2, ∞) is open in R.

3.3.6 Theorem: Let (X,T) be a topological space; then ∅ and X are open sets in X.

Proof. (Exercise for the reader).

3.3.7 Theorem: Let (X,T) be a topological space; then ∅ and X are closed sets in X.

Proof. (Exercise for the reader).

3.3.8 Theorem: Let (X,T) be a topological space; then $\{U_\delta$: $\alpha \in J$ is a collection of closed subsets of X, then $\cap_{\delta\in J} U_\delta$ is closed in X.

Proof. (Exercise for the reader).

3.3.9 Theorem: Let (X,T) be a topological space; then U_α, $\alpha = 1, 2, \ldots, n$, $U_\alpha \subset X_\alpha$, U_α closed, then $\cup_{\alpha}{}^{n}{}_{=1}U_\alpha$ is closed in X.

Proof. (Exercise for the reader).

3.3.10 Theorem: Let (X,T) be a topological space; then

U_α, $\alpha = 1, 2, \ldots, n$, $U_\alpha \subset X_\alpha$, U_α open, then

$\cap_{\alpha=1}^{n} U_\alpha$ is open in X.

Proof. (Exercise for the reader).

3.3.11 Theorem: Let (X,T) be a topological space and

$\{U_\delta : \alpha \in J\}$ a collection of open subsets of X,

then $\cup_{\delta \in J} U_\delta$ is open in X.

Proof. (Exercise for the reader).

Exercise 3-3

1. Prove that in a discrete topological space, each subset is both open and closed.
2. Prove that $\varnothing$, and the set X are both open and closed in any topology on X.

Basis and Subbasis

3.4.0 Basis fof a Topology

3.4.1 Definition Let (X, T_x) be a space. A family $\beta \subset T_x$ is called a basis for T_x if for each $U \in T_x$, U is a union of members of β.

3.4.2 Theorem: Let (X, T_x) be a space and A family $\beta \subseteq T_x$ a basis for T_x, then if p belong to the intersection of two elements B_1 and B_2, then there is a basis element B_3 containing p such that $B_3 \subseteq B_1 \cap B_2$.

Proof: Exercise for the reader.

3.4.3 Theorem: Let (X, T_x) be a space and A family $\beta \subseteq T_x$ a basis for T_x, then for each $p \in X$ there is at least one basis element B containing p.

Proof: exercise for the reader.

3.4.4 Definition: If β is a basis for a topology T_x on X, then the topology generated by β is described as follows: A subset U of X is said to be open in X if for each $p \in U$, there is a basis element $B \subseteq \beta$ such that $p \in B$ and $B \subseteq \beta$.

3.4.5 Theorem: If X is any set, then the collection of all one-point subsets of X is a basis for the discrete topology on X.

Proof: Exercise for the reader.

3.4.6 Theorem: If X is any set and let β be a basis a topology T_x on X, then T_x equals the union of all elements of β.

Proof: Exercise for the reader.

3.4.7 Theorem: If T_x is a topology on X, then T_x is a basis for T_x.

Proof: Exercise for the reader.

3.4.8 Theorem: Let T_x be the discrete topology on X. Then B $= \{\{x\}: x \in X\}$ is a basis for T_x.

Proof: exercise for the reader.

3.4.9 Theorem: Let T_x be the usual topology on the real number R. Then B $= \{(a, b): a, b, \in R\}$ is a basis for T_x.

Proof: Exercise for the reader.

3.4.10 Theorem (a) Let $B \subset T$ be a basis for (X, T). Then A is open ($A \in T$) if and only if for each $x \in A$, there is a $U \in B$ with $x \in U \subset A$.

Proof. ($\Rightarrow$) If A is open, then A $\in$ T, and by Definition 4.11 A is the union of members of B. Hence $x \in A \Rightarrow x \in U \subset A$ and $U \in B$. ($\Leftarrow$) Suppose $x \in A \Rightarrow X \in U_x \in B$ and $U_x \subset A$. Since $B \subset T$, we have that $U_x \in T$; that is, U_x is open. Now we have that $A = \cup_{x \in A} U_x \in T$ Hence, $A \in T$ and A is open.

Finer Topology

3.4.11 Definition a) Let T′ and T′′ be topologies on X. We say that T′ ≤ T′′ or T′ is smaller than T′′ or T′′ is finer than T′ if T′ ⊂ T′′.

b) If p is a property for a topology on X, we say that T′ is the smallest topology with property p if:

(i) T′ has property p and

(ii) If T has a property p, T a topology on X, then T = T′.

3.4.12 Theorem. Let B_1 and B_2 be bases for topologies T_1 and T_2, respectively, on X and T_1 finer than T_2, then for each $p \in X$ and for each $B \in B_1$ with $p \in B$, there is a bases element $B' \in B_2$ so that $p \in B' \subset B$.

Proof. Exercise for the readrer.

3.4.13 Theorem. Let B_1 and B_2 be bases for topologies T_1 and T_2, respectively, on X and each p $\in$ X and for each B $\in B_1$ with p $\in$ B, there is a bases element $B' \in B_2$ so that $p \in B' \subset B$. then T_1 is finer than T_2,

Proof. Exercise for the readrer.

3.4.14 Definition Let $S = \{A_\alpha : \alpha \in J\}$ where $A_\alpha \subset X$ for each $\alpha \in J$. Then there exists a unique smallest topology T_S satisfying $T_S \supset S$. The topology T_S consists of $\emptyset$, X, all finite intersections of the A_α'S, and all arbitrary unions of these finite intersections.

<u>**Subbasis**</u>

Remark 3.4.1 If S and T_S are as in Definitions 3.4.4, then S is called a subbasis for T_S and T_s is said to be generated by S. If one starts with a topology T on X, then a subset $S \subset T$ is called a subbasis for T if $T_S = T$.

3.4.15 Theorem: A family $B = \{B_i: i \in J\}$ of subsets of X is a basis for some topology, T on X if and only if:

(a) $X = \cup_{i \in J} B_i$

(b) For each $p \in X$ and each B_1, $B_j \in B$, $p \in B_i \cap B_j$ there exists a $B_p \in B$ so that $p \in B_p$ and $B_p \subset B_i \cap B_j$.

Proof. (Exercise for the reader).

Equilvalent Basis

3.4.16 Definition. Two basis B and B′ are equivalent if and only if $T_b=T_{B'}$.

3.4.17 Theorem: Two bases B and B′ are equivalent if and only if both of the following hold:

(a) For each $U \in B$ and each $x \in U$ there is a $U' \in B'$ with $x \in U' \subset U$.

(b) For each $U' \in B'$ and each $x \in U'$ there is a $U \in B$ with $x \in U' \subset U$.

Proof. Suppose B′ and B are equivalent basis and $x \in U_x \in B$; by Definition 4.6 $T_B = T_B'$ and by Definition 4.1 $u_x \in T_B$ hence $U_x \in T_{B'}$, take $U_x = U'_x$ hence (a) is true. The proof for (b) is the same; only start with $x \in U'_x \in B'$.

Suppose (a) and (b) hold; then we need only show that $T_B = T_{B'}$. Let $U \in T_B$; then if $U = \emptyset$, clearly $U \in T_{B'}$. If $U \neq \emptyset$, then $p \in U \in T_B$, and U is the union of members of B; that is, $U = \cup_{i\in J}U$, $U_i \in B$, but $p \in B$, but $p \in U \Rightarrow p \in U_p$ for some $p \in J$ and

$U_p \in B$ by (a) there is a $U_p' \in B'$ so that $X \in U_p' \subset U$, now $U = \cup_{U_p \in B}, U_p' \in T_{B'}$. Hence, $T_B \subset T_{B'}$. The proof that $T_{B'} \subset T_B$ is almost the same and is an *exercise for the student*. Hence, $T_B = T_{B'}$ and by Definition 4.6 B is equivalent to B'.

3.4.18 Theorem: Let B be a collection of open sets of X such that for each $U \subset X$, U open and for each $x \in U$, there is a $B_i \in B$ with $x \in B_i \subset U$. then B is a basis for some topology on X.

Proof. (Exercise for the Reader).

3.4.19 Definition. Let X be a space; then B is called a subbasis for the topology, T, of X provided that $T = \{\cup_{n \in N} T_n: \quad T_n = \cap^n{}_{i=1} B_i, \; B_i \in B, \; n = 1, 2, \ldots\}$ is a basis for T.

3.4.20 Theorem: Let X be a set and B a family of subsets of X so that

(a) $B \neq \emptyset$

(b) $\cup_{B_i \in B} B_i = X$.

Then B is a subbasis for some topology for X.

Proof. (Exercise for the Reader).

Exercise Set 3-3.4

1. Let $X = \{a, b, c, d\}$ and let $S = \{\{a\}, \{b, c\}, \{d\}\}$. Then what is T_S?

2. Prove that if B is a basis for T, then B is a subbasis and that the converse is not true.

3. Let d be a metric on X; then show that $B_d = \{ B(x; \epsilon) : x \in X$ and $\epsilon > 0\}$ is a basis for a topology $T_d = T_{(B)d}$ on X.

3.5.0 Closure, Interior, Boundary, Dense Set, Limit Points

In a metric space X and a point $p \in X$ with $A \subset X$, we can say that a point of A is closed to x provided $d(x, A) = 0$. In topological spaces we can find a characterization of closeness.

Closure

3.5.1 Definition. Let X be a space and $S \subset X$. A point $p \in X$ is in the *closure* of S if for each open set U of p, $U \cap S \neq \emptyset$. The close of S is denoted by S^c.

Example 3.5.-1 Let X = R and S = (1, 2,); then S^c = [1, 2].

Example 3.5.-2 Let X = R and S = [3, 5]; then S^c = [3, 5].

3.5.2 **Theorem:** Let X be a space and $S \subset X$; then

If $S \subset F \subset X$ and F is closed, then $S^c \subset F$.

Proof.

Suppose $x \notin F$, then $x \in X\backslash F$ and $X\backslash F$ is open by definition and by Theorem 1-1.8 $X\backslash F \subset X\backslash S$. Hence, $(X\backslash F) \cap S = \emptyset$. Since $X\backslash F$ is open, we have that $x \notin S^c$; then there is an open set U, with $x \in U$ and $U \cap S = \emptyset$ let $F = X\backslash U$. Then F is closed and $S \subset F$. Since $x \in U$, we have that $x \notin F$. Hence, $S^c \subset F$.

3.5.3 Theorem: Let X be a space and $S \subset X$; then

If $x \notin S^c$, then $x \notin F$ for some closed set F, with $S \subset F$.

Proof. Exercise for the reader

3.5.4 Theorem: Let X be a space with $S \subset X$; then $S^c \subset \cap_{i \in J} F_i$ where F_i is closed for each i, and $\{F_i\}_{i \in J}$ is the family of all closed subsets with $S \subset F_i$ for each i.

Proof. (Exercise for reader; Hint: Use Theorem ?.5.2) .

3.5.5 Theorem: Let X be a space with A and B subsets of X, then A is closed if and only if $A = A^c$.

Proof. (Exercise for the reader).

3.5.6 **Theorem**: Let X be a space with A and B subsets of X, then $A^c = (A^c)^c$.

Proof. (Exercise for the reader).

3.5.7 **Theorem**: Let X be a space with A and B subsets of X, then If $A \subset B$ then $A^c \subset B^c$.

Proof. (Exercise for the reader).

3.5.8 **Theorem**: Let X be a space with A and B subsets of X, then $(A \cup B)^c = A^c \cup B^c$.

Proof. (Exercise for the reader).

3.5.9 **Theorem**: Let X be a space with A and B subsets of X, then $(A \cap B)^c \subset A \cap B^c$.

Proof. (Exercise for the reader).

3.5.10 **Theorem**: Let X be a space with A and B subsets of X, then Show by example $(A \cap B)^c \neq A^c \cap B^c$.

Proof. (Exercise for the reader).

3.5.11 **Theorem**: Let X be a space and let $\{F_i\}$ be a family of closed sets; then the following are true:

(a) $\cap_{i \in J} F_i$ is closed

(b) If J is finite, then $\cup_{i \in J} F_i$ is closed.

Proof. (Exercise for the reader).

3.5.12 Theorem: Let (X, T) be a topological space; then for $A \subset X$ and $U \in T$, if $U \cap A^c = \emptyset$, then $U \cap A = \emptyset$.

Proof. (Exercise for the reader).

Remark 3.5.1 In topological spaces, we have found that the closure of a set S is closed, and it is the smallest closed set that contains S. Another important subset related to S is the interior of S. We will find that this set is the largest open set that is a subset of S.

Interior of a Set

3.5.14 Definition. Let X be a space, and $S \subset X$, $p \in S$ is an interior point of S if there is a $U \in T$ with $p \in U \subset S$. The interior of S will be denoted by int(S).

Example 3.5.3 Let X = R, T be the usual topology and S = [1, 4]; then int(S) = (1, 4). Why?

Example 3.5.4 let X = R, T be the usual topology, and S = rational numbers; then int(S) = $\varnothing$. Why?

Example. 3.5.5 Let X = {a, b, c}, T = {$\varnothing$, x, {a, b}}, and let S = {a, b,}; then int(S) = {a, b}. Why?

3.5.15 Theorem: Let (X, T) be a space, $S \subset X$; then int(S) = $X \backslash (X \backslash S)^c$

Proof. (Exercise for the reader).

3.5.16 Theorem: Let (X, T) be a space and $S \in T$; then int(S) = S.

Proof. $p \in$ int(S); then there is a $U \in T$ with $p \in U \subset S$, and we have int(S) $\subset$ S. Suppose $P \in S$, since $S \subset T$ and $p \in S \subset S$ now we have $p \in$ int(S) and $S \subset$ int(S). Thus int(S) = S.

3.5.17 Corollary. Let (X, d) be a space; $s \subset X$ then $\text{int}(S) = \cup \{U: U \in d \text{ and } U \subset S\}$.

Boundary of A Set

3.5.18 Definition. Let (X, T) be a space; then the boundary of a set $S \subset X$ (denoted by Bdry(A)) = $S^{c}n\ (x|S)^{c}$

3.5.19 Theorem: Let (X, T) be a space and $S \subset X$; then $S^{c} = S \cup \text{Bdry}(S)$.

Proof. Clearly, $S \subset S^{c}$ and $\text{Bdry}(S) = S^{c} \cap (x|S)^{c} \subset S^{c}$, hence $S \cup \text{Bdry}(S) \subset S^{c}$. Also, $p \in S^{c}$; then $p \in S$ or $p \notin S$. If $p \notin S$, then $p \in (X|S)^{c}$; hence, $p \in S^{c}\, n(X|S)^{c} = \text{Bdry}(S)$, and $p \in S^{c} \cap (X|S)^{c} = \text{Bdry}(S)$, and $p \in S \cup \text{Bdry}(S)$. If $P \in S$, then clearly $p \in S \cup \text{Bdry}(S)$. Hence, $S^{c} \subset S \cup \text{Bdry}(S)$ and $S^{c} = S \cup \text{Bdry}(S)$.

3.5.20 Theorem: Let (X, T) be a space $S \subset X$; then $S^{c} = S \cup S'$.

Proof. (Exercise for the reader).

Dense Set

3.5.21 Definition. Let (X, T) be a space $D \subset X$; then D is dense in X if and only if $D^{c} = X$.

Example 3.5.6. Let X =n {a, b, c,} and d = {∅, X, {a, c}} and D = {c}, then D^c = X; hence, D is dense in X.

3.5.22 Theorem: Let (X, T) be a space and D ⊂ X be dense in X with S a basis for T; then for each ∅ ≠ U ∈ S, U ∩ D ≠ ∅.

Proof. (Exercise for the reader).

3.5.23 Corollary. Let (X, T) be a space and ∅ ≠ U ∈ T with D ⊂ X and D dense in X; then U ∩ D ≠ ∅.

Proof. (Exercise for the reader).

Exercise Set 3.5 (Optional)

1. Let (X, T) be a space and S ⊂ X. Prove the following:

   (a) Bdry(S) is closed.

   (b) Bdry(S) ∩ int(S) = ∅.

   (c) Bdry(S) ∩ int(X\S) = ∅.

   (d) Int(S) ∩ int(X\S) = ∅.

   (e) Bdry(S) = ∅ if and only if S ∈ T.

Limit Points

3.5.24 Definition. A point p is a limit point of A ⊂ X, (X, T) a topological space then for each U ∈ T, p ∈ U, then U ∩ A ≠ ∅.

Example 3.5.3. Let X = {0, 1, 2, 3}, d = {∅, X {0, 1}, {1, 2, 3}} then 1 is a limit point of {0, 1}.

Exercise Set 3.5

1. Let (X, T) be a space and let $S = \cup\ \{U_d\text{: } d \in J,\ X\backslash U_d \in T\}$, prove or disprove that S is closed. Let $Q = \cap_{i\in J}\{U_i\text{: } U_i \in d\}$, prove or disprove that Q is close.
2. Bdry(S) ⊂ S′, S′ ⊂ S′ ⊂ Bdry(S) for each subset S of space X.
3. Let X = R and d be the usual topology and S = rational numbers. What is S′? What is T? What is Bdry(S)?
4. Let (X, T) be a space, S ⊂ X. Prove that Bdry(S) ⊂ S if and only if $S = S^c$, and Bdry(S) ⊂ X\S if and only if S ∈ T.

3.6.0 Continuity

In the event that f is a function from a topological space (X, T) to a topological space (Y, T′), we shall write f:X→Y, as we indicated in Chapter 2 for metric spaces. Recall that a function f:R→R is continuous at p ∈ R if and only if for each ∈ > 0 there is a δ > 0 so that |f(x) - f(p)| < ∈ whenever |x - p|

$< \delta$ or $f(x) \in (f(p) - \epsilon, f(p) + \epsilon)$ whenever $x \in (p - \delta, p + \delta)$. This implies the following definition.

Continuity and Topological Spaces

3.6.1 Definition. Let (X, T) and (Y, T') be spaces; then $f:X \to Y$ is continuous at $p \in X$ if and only if for each $U \in T'$ with $f(p) \in U$; then there is a $V \in T$ with $p \in V$ and $f(V) \subset U$. A function $f:X \to Y$ is said to be continuous if it is continuous at each $p \in X$.

Example 3.6.1 Let $X = \{a, b, c\}$, $d = \{\{a\}, \{b\}, \{c\}, \{a,b\}, \{b,c\}\ \emptyset, X\}$ and $Y = \{0,1\}$, $d' = \{\emptyset, Y, P\{0\}\{1\}\}$; then (X, d) and Y, d') are spaces. Also, let $f:X \to Y$ by $f(a) = f(b) = 0$, $f(c) = 1$; then f is continuous for $f(a) \in \{0\}$ and $a \in \{a,b\}$, $a \in \{a\}$. Also, $f(\{a,b\}) = f(\{a\}) = \{0\} \subset \{c\}$. Also $b \in \{a, b\}$, $b \in \{b\}$ and $f(\{a, b\}) = f(\{b\}) = \{0\}$, $f(c) \in \{1\}$, $c \in \{c\}$ and $f(\{c\}) \subset \{1\}$.

3.6.2 Theorem: Let (X, T), (Y, T') be topological spaces $f:X \to Y$; then the following are equivalent (T.A.E).

(a) f is continuous

(b) $U \in T'$ then $f^{-1}(U) \in T$

(c) $(Y \setminus U) \in T'$ then $f^{-1}(U)$ is closed in X

(d) $\forall A \subset X$, $f(A^c) \subset [f(A)]^c$

Proof. a)$\Rightarrow$b) Let $U \in T'$, take $p \in f^{-1}(U)$, then $f(p) \in U$; that is, U is an open set with $f(p) \in U$. Since f is continuous, there is a $V \in d$ with $p \in V$ and $f(V) \subset U$. Hence, $p \in V \subset f^{-1}(f(V)) \subset f^{-1}(U)$. Thus, $f^{-1}(U)$ is open and, hence, $f^{-1}(U) \in d$.

b)$\Rightarrow$c) Since $Y \setminus U \in T'$, U is closed in Y, by (b) $f^{-1}(Y \setminus U) \in T$ and by Theorem 1.4.4 $f(Y \setminus U) = f^{-1}(y) \setminus f^{-1}(U) = X \setminus f^{-1}(U)$. Hence, $X \setminus f^{-1}(U) \in T$; that is, $X \setminus f^{-1}(U)$ is open in X and $f^{-1}(U)$ is closed in X.

c)$\Rightarrow$d) Note that $A \subset f^{-1}(f(A)) \subset f^{-1}((f(A))^c$ and by C $f^{-1}((f(A))^c)$ is closed in X. Thus, A^c c $f^{-1}((f(A))^c) \subset f^{-1}(f(A^c)) \subset f^{-1}((f(A))^c)$. Thus $f(A^c) \subset (f(A))^c$.

d)$\Rightarrow$a) Exercise For The Reader

3.6.3 Theorem: Let (X, T) and (X, T') be spaces and $f:X \to X$ be defined by $f(p) = x$ for each $p \in X$

(a) f is continuous if and only if $T' \subset T$

(b) f is continuous if and only if $T \subset T'$

Proof. (Exercise for the reader;

3.6.4 Theorem: Let (X, T), (Y, T'), (Z, T'') be spaces and $f:X \to Y$, $g:Y \to Z$ be continuous. Then $g \circ f:X \to Z$ is continuous.

Proof. (Exercise for the reader; Hint: Use Theorem 3.6.2).

<u>Open & Closed Map</u>

3.6.5 Definition. Let (X, d), (Y, d') be spaces; then a map $f:X \to Y$ is an open map if and only if for each open set $A \subset X$, $f(A)$ is open in Y.

3.6.6 Definition. Let (X, d), (Y, d') be spaces; then a map $f:X \to Y$ is a Closed map if and only if for each closed set $A \subset X$, $f(A)$ is closed in Y.

Example 3.6.1 Let $X = Y = R$ and let $f:X \to Y$ by $f(x) = x$ for each $x \in X$; then f is both an open and closed map.

Homeomorphism

3.6.7 Definition. Let (X, T), (Y, T′) be spaces; then a map $f:X \to Y$ is a homeomorphism if and only if f is bijective and f is both open and closed. If such $f:x \to y$ exists, then space X is homeomorphic to space Y.

Topological Property

3.6.8 Definition. A property that is preserved by a topological space Y is also preserved by every topological space that is homeomorphic to Y and is called an intrinsic qualitative property or a topological property of the space Y.

Note: For some examples, see Remarks; also, see examples at the first part of this chapter.

Example 3.6.2 Let X = {a, b, c} and let T = {∅, X, {a, b}, {c}}. Let Y = {d, e, P}, and let T′ = {∅, Y, {d, e}, {P} and define $f:X \to Y$ by f(a) = d, f(b) = e, f(c) =p; then f is a homeomorphism. Thus, (X, T) and (Y, T′) are homeomorphic.

Remarks: 3.6.1: We have made mathematically precise the notion of stretching and bending without tearing (intrinsic qualitative property) that we referred to at the beginning of this chapter. The mappings that stretch and bend their domain into their ranges without tearing are called **homeomorphism.**

It should be clear by now that the simplest way to show that two spaces are not homeomorphic to find a topological property of one space that the other space fails to have. A property that is defined only in terms of the points and open sets of the space are clearly topological properties, because a homeomorphism is a one-to-one correspondence between the points of the two spaces and a one-to-one correspondence between the open sets of the two spaces.

Example 3.6.3 Let X = (a, b, c} and Y = (d, e, f}; also let T_x = {∅, X, {a, b}, {c}. T_Y = {∅, Y, {d, c}, {d, f}, {e, f}, {d}, {e}, {f}}. Then space X is not homeomorphic to space Y. Why?

Exercise Set 3.6

1. Let X = {a, b, c}, and let T_x = {∅, X, {a, b}, {c}}.

    (a) Give the basis B for T_x.

    (b) Give the subbasis B′ for Tx where B′ ⊂ B.

(c) Give an example of a nonconstant, continuous function of X into R.

(d) Give an example of a nonconstant function of X into R that is not continuous.

2. If (X, T) is a space, then following and true:

(a) S open in X, $A \subset X$ then $(S \cap A)^c = (S \cap A^c)^c$

(b) Let $\{s_i\}^n_{i \in 1}$, S open in X for each i and $(S_i)^c = X$ for each i, then $\cap^n_{i=1} S_i = X$.

3. Let (X, T) be a space and B a basis for d; then B is dense in X.

4. Let (X, T), (Y, T′) be spaces and B, S be basis and subbasis for T and T′, respectively.

(a) A map $f: Y \to X$ is continuous if and only if $f^{-1}(B_i)$ is open in Y for each $B_i \in B$.

(b) A map $g: X \to Y$ is continuous if and only if $f^{-1}(S_i)$ is open in X for each $S_i \in S$.

3.7.0 Product Spaces

In this section we will define the product topology for any collection of spaces and study some basic properties of product spaces. We first consider the finite case, that is, a finite product of spaces.

Suppose X_1 and X_2 are spaces; then from Chapter 1 we have that if U_1 and U_2 are open in X_1 and X_2, respectively, $p^{-1}(U_1) = U_1 \times X_2$ and $p_2^{-1}(U_2) = X_1 \times U_2$, are each open in $X_1 \times X_2$. Note that p_1, p_2 are the projections as defined in Chapter 1.

Box Topology

3.7.1 Definition. Let X_1, X_2 be spaces; the box topology of $X_1 \times X_2$ is the topology with basis $B = \{U_1, \times U_2 : U_1$ and U_2 open in X_1 and X_2 respectively$\}$.

Remark 3.6.2 If U_1, U_2 are open in X and V_1, V_2 are open in Y, then $U_1 \times V_1 \cap U_2 \times V_2 = U_1 \cap U_2, \times V_1 \cap V_2$ (See Figure 11 below).

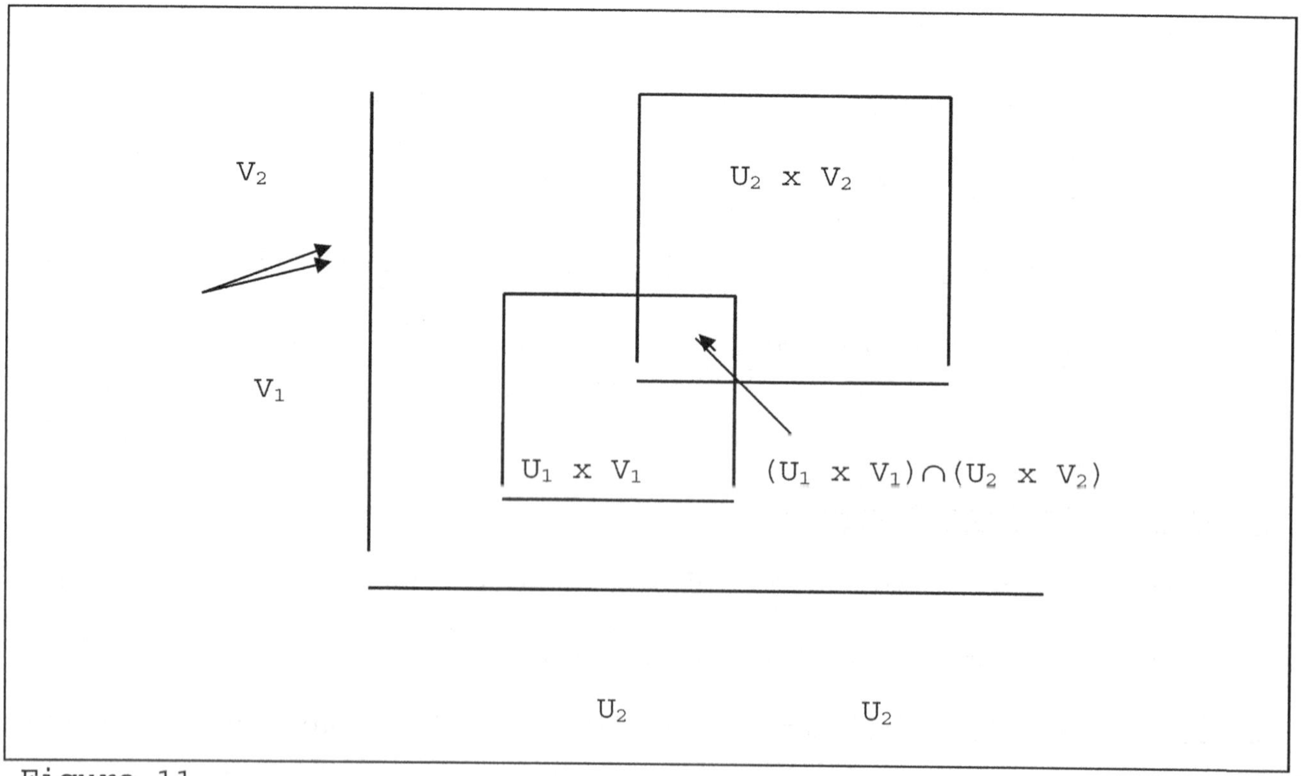

Figure 11

This topology is called the box topology for X x Y.

3.7.2 Theorem: Let X and Y be spaces and B = $\{p_1^{-1}(U):$ U is open in X$\}$ $\cup\{P^{-1}(V):$ V open in Y$\}$; for the box topology on X x Y.

Proof. Let T be the box topology on X x Y, and T′ be the topology generated by B. Since every member of B is a member of T, so do arbitrary unions of finite intersections of elements of B. Hence, d′ $\subset$ d. On the other hand, every basis element U x V of the topology T is a finite intersection of elements of B, since U x V = $p_1^{-1}(U) \cap p_2^{-1}(V)$. Thus, T $\subset$ T′ and, hence, T = T′, (See Figure 12 below).

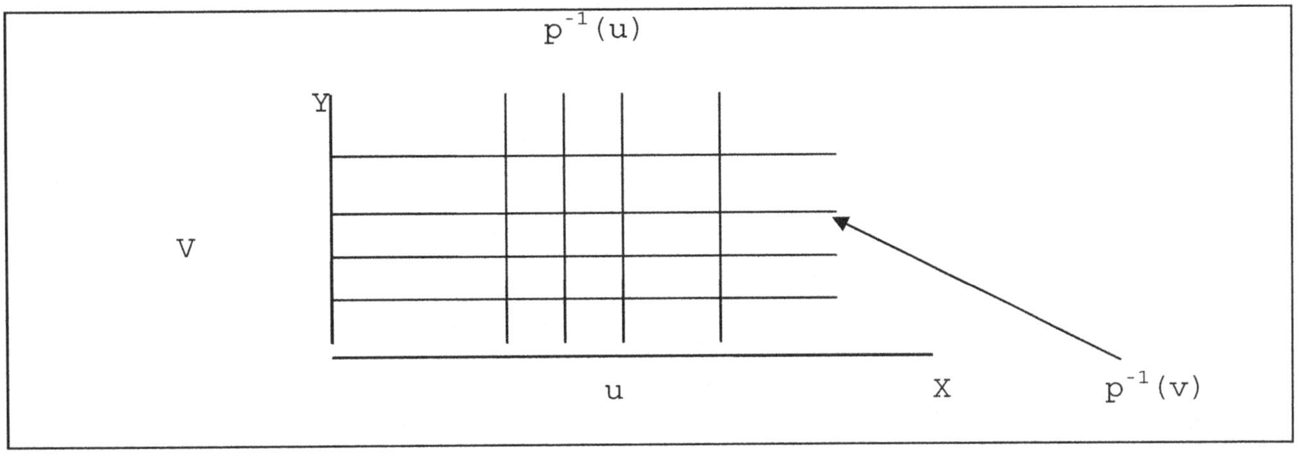

Figure 12

<u>Product Topology</u>

3.7.3 Definition. Let $\{X_i:$ i $\in$ J) be a collection of spaces. Let T_i be the topology for X_i for each i $\in$ J. The product topology in $\pi_{i\in J}X_i$ is the topology with subbasis B = $\{<U_i> : <U_i> = p_i^{-1}(U_i)$ where

$U_i \in T_i$ and i over all elements of J}. The box topology in $\pi_{i \in J}X$ is the topology with basis B = $\{\pi_{i \in J}U_i$: where U_i is open in X_i for each i ∈ J.

3.7.4 Theorem: If J is finite; then box topology coincides with the product topology as defined in Definition 7.3.

Proof. (Exercise for the reader; Hint: a proof is only needed for n = 2).

3.7.5 Theorem: Let $\pi_{i \in J}X_i$ be the product of spaces with the product topology; then If $A_i \subset X_i$ for each i ∈ J, then $(\Pi_{iej}A_i)^c = \Pi_{iej}(A_i)^c$

Proof. (Exercise for the reader).

3.8.0 Subspaces

In the case of the metric space (X,d) we obtained the subspace by placing a restriction on the metric d to the subset A ⊂ X. Also, we obtained the fact a set U open in A, then U = A ∩ U′ for some set U′ of X. Thus, we have the following definition.

Topological Subspace

3.8.1 Definition. Let (X, T) be a space and $A \subset X$. Let $T' = \{U \cap A: U \in d\}$. T' is called the relative topology for A induced by T. Also, (A, T') is called a subspace of (X, T).

3.8.2 Theorem: Let T' be defined as in Definition 3.8.1. Show that (A, T') is a topological space.

Proof. (Exercise for the reader).

Example 3.6.4 Let $X = \{a, b, c\}$, $d = \{\emptyset, X, \{a, b\}\}$ and $A = \{b, c,\}$ with $T' = \{U \cap A: U \in d\} = \{\emptyset, A, \{b\}\}$.

Example 3.6.5 Let $X = R^2$ and T the usual topology on R^2 and $A = \{P(x, y): x^2 \div y^2 = 1\}$. What is $T' = \{U \cap : U \in d\}$?

Remarks 3.6.4 If A is as subspace of X and $B \subset A$, then B^c in A may not be the same as B^c in X. We now express in relation between B^c and A and B^c in X.

Exercises Set 3-6

1. Is Theorem 8.4 true if the conditions A-open and A-closed are dropped?

2. Let X = R and A = rational numbers with T the usual topology on R. What is T_A? What are the closed and open sets in A?

3. Let A be a subspace of X and X a subspace of Y. Show that A is a subspace of Y.

Chapter 4

Various Types of Topological Spaces

In this chapter we shall give some attention to some interested and desirable properties that result when special restrictions are placed on topological spaces. In most cases the spaces we study in calculus, analysis, and other areas of mathematics possess one or more of these properties. The first of these restrictions that we shall consider will be what we call the separation properties.

4.1.0 Separation Properties

Here we consider to what extent one can separate closed sets and sets of the form $\{x\}$ with open sets.

4.1.1 Definition. A space x is called a T_o-space if and only if for each x, y, $\in X$, $x \neq y$, either there is an open set U with $x \in U$, $y \notin U$ or there exists an open set V with $y \in V$ and $x \notin V$.

Example 4.1.1 Let $X = \{a, b, c\}$ and $T_x = \{\varnothing, X, \{b\}, \{c\}, \{b, c\}\}$; then (X, T_x) is a T_0-space.

T_1-Space

4.1.2 Definition. A space X is called a T_1-space if and only if for each $x, y \in X$, $x \neq y$, there are open sets U, V so that $x \in U$, $y \notin U$, and $x \notin V$, $y \in V$.

Example 4.1.2 Let $X = R$, T_x be the usual topology; then (X, T_x) is the T_1-space.

4.1.3 Theorem: Let X be a space; then X is T_1 if and only if $\{x\}$ is closed for each $x \in X$.

Proof. (Exercise for the reader).

<u>T_2-Space</u>

4.1.4 Definition. A space X is called T_2-space (Hausdoff Space) if and only if for each $x, y \in X$, $x \neq y$, these exist open sets u,v so that $x \in u$, $y \in v$ and $u \cap v = \varnothing$.

Example 4.1.3 (See Example 4.1.2).

<u>T_3-Space</u>

4.1.5 Definition. A space X is called T_3-space if and only if for each $x \in X$, and each closed set $F \subset V$, there exist open sets u,v with $x \in u$, $F \subset v$ and $u \cap v=\varnothing$.

Example 4.1.4 Let $K = \{a, b, c\}$ and let $T_x = \{\varnothing, X, \{a\}, \{b, c\}\}$; then (X, T_x) is a T_3-space (See Figure 1 below).

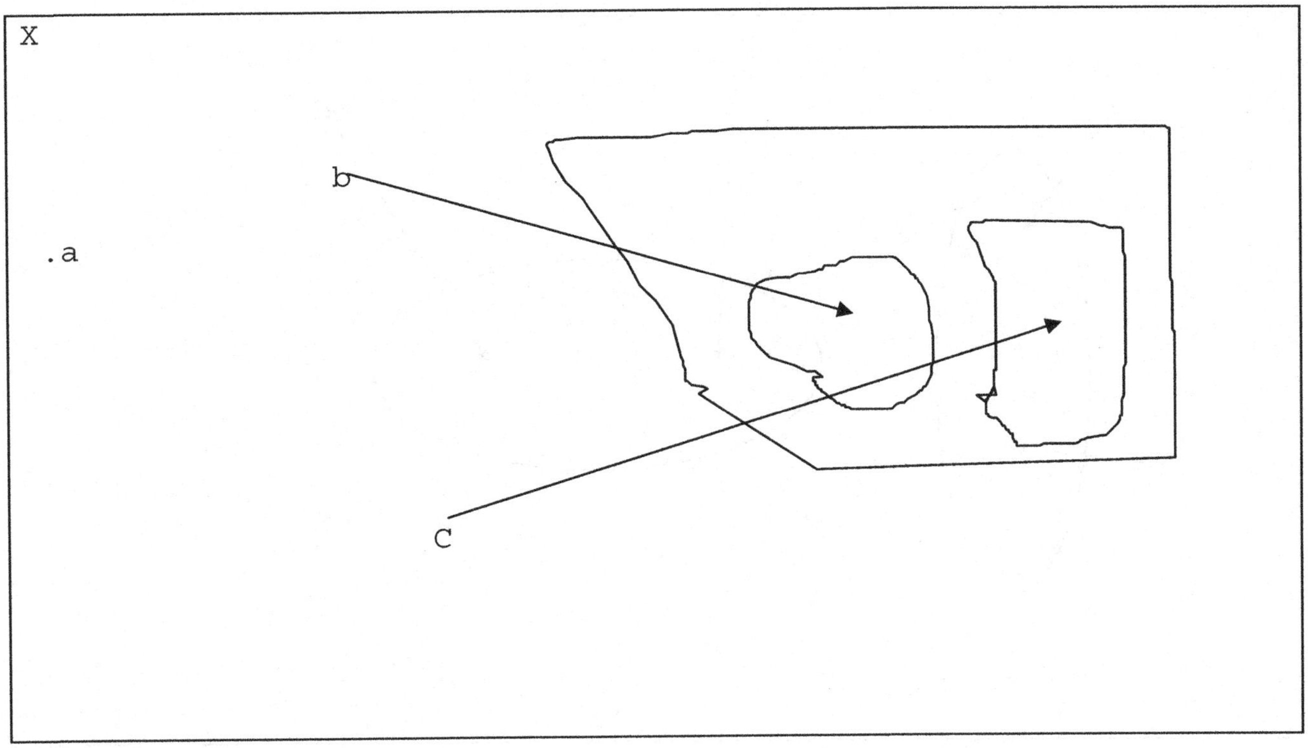

Figure 1

<u>T_4-Space</u>

4.1.6 Definition. A space X is called a T_4-space if and only if for each pair of disjoint closed sets $A, B \subset X$, there exist open sets U, V with $A \subset U$, $B \subset V$, and $U \cap V = \varnothing$.

Example 4.1.5 Let $X = \{a, b, c, d\}$ and let $T = \{\varnothing, X, \{a, b,\}, \{c, d\}\}$; then (X, T) is a T_4-space (See Figure 2).

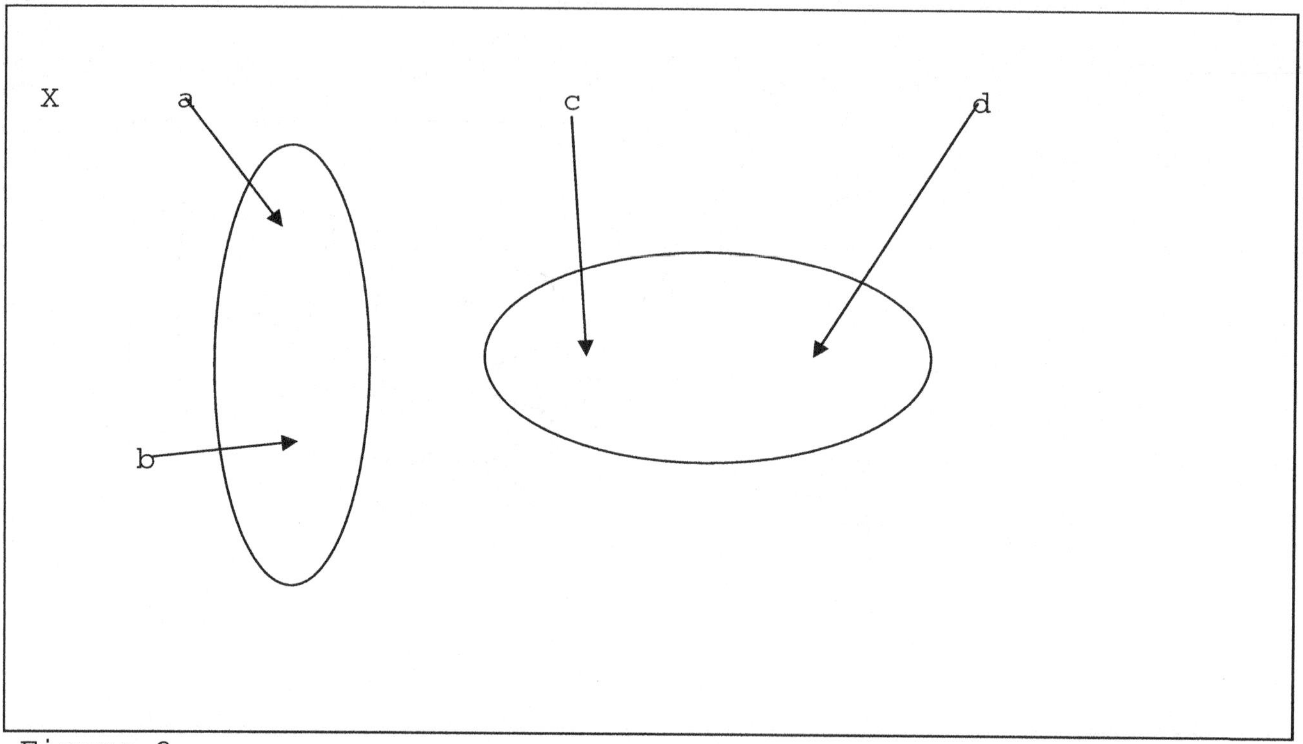

Figure 2

4.1.7 Theorem: Let X be a finite T_1-space; then d_x I the discrete topology.

Proof. (Exercise for the reader; Hint: See Theorem 4.1.3).

<u>Regular and Normal Spaces</u>

4.1.8 Definition.

(a) A space X is called a regular space if and only if X is both T_1 and T_3.

(b) A space X is called a normal space if and only if x is both a T_1 and T_4 space.

Remark 4.1.2 From the figure below we can conclude that a normal space is a regular space, a regular space is a T_2-space, a T_2-space is a T_1-space, and a T_1-space is a T_0-space.

Note on terminology: The notation T1-space is due to Alexandroff and Hopf and differs somewhat from what we have done here. The difference si slight and amounts to the following: What we call a regular space, they call a T_3-space; What we call a normal space, they call a T_4-space. Some writers follow Alexandroff and Hopf, and some do as we have done; consequently, the reader should be aware of the alternative meanings of the terms "T_3-space" and "T_4-space."

4.1.9 Theorem:

(a) A space X is regular if and only if X is a T_1-space and for each $x \in X$, U open in X with $x \in U$, there exists an open set V with $x \in V \subset V^c \subset U$.

(b) A space X is normal if and only if X is a T_1-space, and for each closed set $F \subset X$ and each open set U with $F \subset U$, there exists an open set V such that $F \subset V \subset V^c \subset U$.

Proof.

(a) First let X be regular, $x \in X$ and U open with $x \in U$; then there is a W open with $x \in W \subset U$. $X \setminus W$ is closed, and $x \in X \setminus$

W; thus, since X is T3 there exist open sets V, V′ with $x \in V$, $X \setminus W \subset V'$ and $V \cap V' = \varnothing$). Thus, we have that $x \in V \subset V^c \subset U$, because $W' \subset W \subset U$ and W′ is closed with $V \subset W'$.

Conversely, let X satisfy the given conditions; since X is T_1, we need only show that X is T_3. Let $x \in X$ and $F \subset X$, F closed, $x \notin F$; then $x \in U = X \setminus F$ and U is open. Take V open with $x \in V$; then by the conditions we now take $x \in V \subset V^c \subset U$. Take $W = X \setminus V^c$; then W is open and $W \cap V = \varnothing$ and $F = X \setminus U \subset X \setminus V^c = W$. Hence, X is T_3, and the proof is complete.

(b) (Exercise for the reader; Hint: See the proof of (a)).

4.1.10 Theorem: Normality$\Rightarrow$Regularity$\Rightarrow T_2 \Rightarrow T_1 \Rightarrow T_0$.

Proof. <u>Normality$\Rightarrow$Regularity</u>. Let K be a normal space; then K is T_1; thus, we need only show that X is a T_3-space. Suppose F is a closed set in X, and $x \notin F$, but by Theorem 4.1.3 $\{x\}$ is closed and by normality there exist open sets U, V with $\{x\} \subset U$ and $F \subset V$ and $U \cap V = \varnothing$; thus, X is T_3 and hence, regular.

<u>Regularity$\Rightarrow T_2$</u>. Let $x, y \in X$, $x \neq y$; since X is T_1 $\{x\}$ is closed and since X is T_3, there exist open sets U, V with $y \in U$ and $\{x\} \subset V$ and $U \cap V = \varnothing$; hence, X is a T_2-space.

$\underline{T_2 \Rightarrow T_1}$. Let $x, y \in X$, $x \neq y$; then since X is T_2, there exist open sets U, V with $x \in U$, $y \in V$ and $U \cap V = \emptyset$. Therefore, $x \notin V$ and $y \notin U$, so X is T_1.

$\underline{T_1 \Rightarrow T_0}$. Let $x, y, \in X$, $x \neq y$; then since X is T_1, there exist open sets U, V, with $x \in U$, $y \notin U$ and $y \in V$, $x \notin V$. Therefore $x \in U$ and $y \notin U$ or $x \in V$ and $y \notin V$. Hence, X is a T_0-space.

4.1.11 Theorem: Every metric space is a Hausdoff Space (T_2-space).

Proof. (Exercise for the reader).

Exercise Set 4-1.

1. Let X be a T_1-space, x a limit point of $A \subset X$; then for each open set U, $x \in U$, $U \cap A$ is infinite.

2. Let X be a T_1-space, $\leftarrow A \subset X$, A finite; then A has no limit points.

3. Show that a T_3-space need not be a T_2-space.

4. Show that a T_1-space need not be a T_2-space.

5. Show that every metric space is normal.

6. A finite T_1-space has no limit points.

4.2.0 Connected

Here we consider the property of connectedness. Intuitively, we may think of a space as being connected if it is not two or more pieces. Clearly, the real line R is connected, and the real plane R^2 is connected. $R\setminus\{1\}$ si not one piece, and we would want this not to be connected. This is the case for $R \setminus \{1\}$ is the union of $U = \{x:x > 1\}$, $U' = \{x: x < 1\}$ and $U \cap U' = \varnothing$. It certainly is reasonable for us to make our definition as follows:

Separated and Connected Sets

4.2.1 Definition: Let X be a topological space $A, B \subset X$, A, B nonempty; then

(a) A and B are separated if and only if $A \cap B^c = A^c \cap B = \varnothing$.

(b) X is connected provided X is not the union of two separated sets.

Example 4.2.1 Let $X = \{0, 1\}$ and let $T' = \{\emptyset, X, \{0\}, \{1\}\}$; then X is not connected because $X = \{0\} \cup \{1\}$. (See Figure 3).

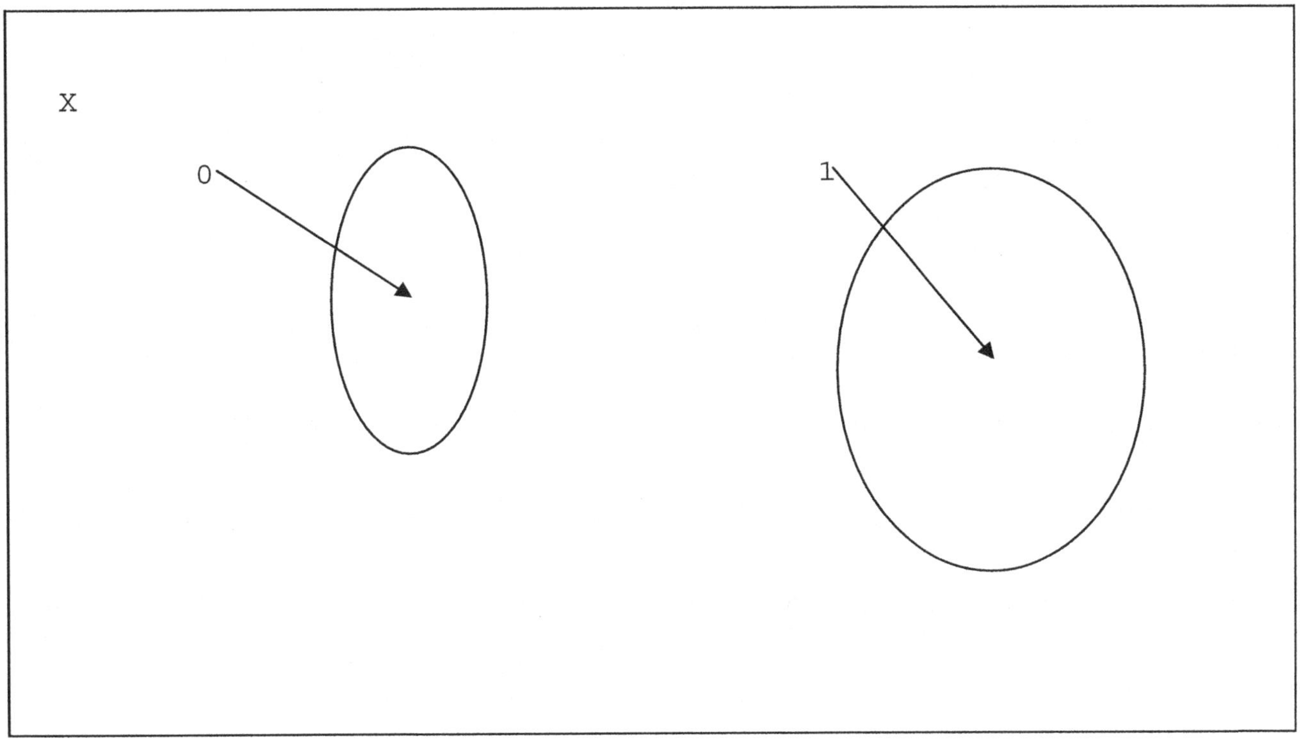

Figure 3

4.2.2 Theorem: Let X be a connected space, then X is not the union of two disjoint, nonempty open subsets.

Proof. (Exercise for the reader).

4.2.3 Theorem: If X is not the union of two disjoint, nonempty open subsets, then X is connected.

Proof. (Exercise for the reader).

4.2.4 Theorem: Let X be a connected space, then X is not the union of two disjoint, nonempty closed subsets.

Proof. (Exercise for the reader).

4.2.5 Theorem: If X is not the union of two disjoint, nonempty closed sub sets, then X is connected.

Proof. (Exercise for the reader).

4.2.6 Theorem: Let X be a connected space, then the only subsets of X both open and closed are $\varnothing$ and X.

Proof. (Exercise for the reader).

4.2.7 Theorem: If the only subsets of X both open and closed are $\varnothing$ and X, then X is connected.

Proof. (Exercise for the reader).

4.2.8 Theorem: Let X be connected and let $f: X \to Y$ be continuous; then $f(X)$ is connected in Y.

Proof. (Exercise for the reader).

4.2.9 Theorem: Let X be connected space, and let $A_i \subset X$ for each $i \in J$ with A_i connected for each $i \in J$ and $\cap_{i \in J} A_i \neq \emptyset$, then $\cup_{i \in J} A_j$ is conntected.

Proof. (Exercise for the reader).

4.2.10 Theorem: Let $A \subset X$ be connected so that for any set B with $A \subset B \subset A^c$, then B is connected.

Proof. (Exercise for the reader).

4.2.11 Theorem: The only connected subsets of R with more than one point are R and intervals (open, closed, half open).

Proof. We first show that X is connected if and only if X is an interval. Suppose X is connected; then X is an interval because if X is not an interval, then for some a, b, $\in$ X, there is a, c $\in$ X such that $a < c < b$, and we have $U = X \cap \{p: p > c\}$, $V = X \cap \{p: \quad p > c\}$ is a separation of X; hence, X is not connected. Thus, X is an interval.

Conversely, suppose X is an interval; then X is connected because if X is not connected, then $X = A \cup B$ where $A \cap B \neq \varnothing$, A, B, open, and there exist $a \in A$, $b \in B$ with $a < b$ (relabeling if necessary). Define $A_y = \{y: [a,y) \subset A\}$, and $g = \text{Sup}\{A_y\}$; then $g \leq b$ and $g \in X$. Also, $g \in A^c{}_y \subset A^c = A$, since $A = X \setminus B$ is closed in X; hence, $y \in A$. A is also open in X and by Theorem 3.4.2, and since X is an interval, there is a $U = (g-\epsilon, g+\epsilon) \subset A$, which means that there is a $p \in A$, $p > g$, contradiction; thus, X is connected.

Now we show that R is connected. Since $f: (0,1) \rightarrow R$, defined by

$$x \longrightarrow \frac{(2x-1)}{(x-1)}$$

is a homeomorphism, and since (0,1) is connected, by Theorem 4.2.3 R is connected.

4.2.12 Theorem: Let X be a space with subsets A, B a separation of X; also, let $C \subset X$ be connected, then $C \subset B$ or $C \subset A$.

Proof. Assume that the conclusion is false, that is, $A \cap C \neq \varnothing$ and $B \cap C \neq \varnothing$; then $C = (A \cap C) \cup (B \cap C)$, and $(A \cap C) \cap (B \cap C) \neq \varnothing$. Hence, $\{A \cap C, B \cap C\}$ is a separation of C; this means that C is not connected and is a contradiction. Thus, $A \cap C \neq \emptyset$, which means $C \subset A$ or $C \subset B$.

4.2.13 Theorem: Let X be a space and S a collection of connected subsets of X such that for each C_1 , $C_2 \in S$, $C_1 \cap C_2 \neq \emptyset$, then $\cup S$ is connected.

Proof. (Exercise for the reader).

4.2.14 Lemma: Let X and Y be spaces; then X and Y are connected if X x Y is connected.

Proof. (Exercise for the reader).

4.2.15 Theorem: Let $\{X_i\}_{i \in J}$ be a family of spaces; If $\pi_{i \in J} X_i$ is connected and $\pi_{i \in J} X_i$ has the product topology, then X_i is connected for each $i \in J$.

Proof. (Exercise for the reader).

4.2.16 Definition: Let X be a space, and $y \in X$. The component $C_{(y)}$ of $y \in X$ is the union of all connected subsets of X containing y.

4.5.2 Definition: Let X be a set and (P_i), $i \in J$ be a collection
of non-empty subsets of X with

a) $X = \bigcup_{i \in J} P_i$;

b) and if $\alpha, \beta \in J$, $\alpha \neq \beta$, then $P_\alpha \neq P_\beta$,

then (P_i), $i \in J$ if a partition of X.

4.2.19 Theorem: Let X be a space, then each $C_{(y)}$ is a maximum connected subset in X for each $y \in X$.

Proof: Exercise for the reader.

4.2.20 Theorem: Let X be a space, then we have that each $C_{(y)}$ is closed X for each $y \in X$.

Proof: Since $C_{(y)}$ is connected by Theorem 4.2.5 $C_{(y)}$ and $C_{(y)} \subset$ closure of $C_{(y)}$. Also by Theorem 4.2.19 we have $C_{(y)} \subset$ closure of $C_{(y)} \subset C_{(y)}$. Thus $C_{(y)}$ = closure of $C_{(y)}$ and $C_{(y)}$ is closed.

4.2.21 Theorem: Let X be a space, then we have that the set of all distinct components in X form a partition of X.

Proof: Exercise for the reader.

4.2.22 Theorem: Let X be a space, then we have that $C_{(y)}$ need not be open in X for each $y \in X$.

Proof: Exercise for the reader.

4.2.23 Theorem: Let X be a space and B connected in X, then the closure of B is closed in X.

Proof: Exercise for the reader.

Remark 4.2.3 It is clear from Theorem 4.2.4 that $C_{(y)}$ is connected for each point y in a space Y. The number of components provides a rough indication as to how "disconnected" a space is.

Examples 4.2.2

Consider the subspace Q of rational numbers of the real numbers R,. The component of each $y \in Q$ is the point y.

Example 4.2.3

Consider the real numbers R. The component of each point y in R is the space R.

4.2.24 Theorem: Let A_x be connected set for each $x \in \propto$ and $\cap_{x\in\propto} A_x \neq \emptyset$, then $\cup_{x\in\propto} A_x$ is connected.

Proof: Exercise for the reader.

Remark 4.2.4 As a result of Theorem 4.32.19 we may ask the question "What is a sufficient condition for a component in a space to be open?" The answer is the space must be locally connected.

4.3.0 Locally Connected

4.3.0 Definition. A space X is **locally connected** at a point p in X if and only if for every neighborhood U of p contains a connected neighborhood of p. The space X is locally connected if it is locally connected at each of its points.

4.5.2 Theorem: The real line R is locally connected.

Proof: Exercise for the reader.

4.5.3 Theorem: The subspace $[2,3) \cup (3,5]$ of R is not connected, but it is locally connected.

Proof: Exercise for the reader.

4.3.3 Theorem: Let X is locally connected space, then each component in X is open.

Proof. (Exercise for the reader).

4.3.4 **Theorem:** Let X be a space so that each component in X is open, then X is locally connected.

Proof. (Exercise for the reader).

4.5.2 Theorem: Let space X be is locally connected at a Point $P \in X$, then there is a basis for the Neighborhoods at p composed of connected subsets of X.

Proof: Exercise for the reader.

4.5.3 Theorem: If for each point $P \in X$, there is a basis for the Neighborhoods at p composed of connected

subsets of X, then X is locally connected at

the point $P \in X$.

Proof: Exercise for the reader.

4.4.0 Path Connected

We attack one more type of connectedness, which is of considerable utility in topology. The motivation is that some of the spaces that are familiar have the property that points may be joined by paths, some examples are the real line, plane, and the Euclidean space $R^3 = \{(a, b, c): a, b, c \text{ real numbers}\}$. In the calculus of the plane, one often discusses a curve in terms of a parametric equation,

$x = F(t)$,

$y = g(t)$,

$$z = h(t).$$

where the functions f, g, and h are at least continuous and possibly differential over some common interval [a, b] as their domain; thus $F(t) = (f(t), g(t), h(t))$ defines a continuous function $F: [a, b] \rightarrow R^3$. The path is the image of [a, b] under F. That is, the path is the curve connected the two points $F(a) = (f(a), g(a), h(a))$ to $F(b) = (f(b), g(b), h(b))$. The interval [a, b] may be restricted to [0,1], for using the homeomorphism $\Psi: [0,1] \rightarrow [a, b]$, one can show the required function $F: [a, b] \rightarrow R^3$ exist if and only if a corresponding $G = F \circ \Psi: [0, 1] \rightarrow R^3$ exist. The above remarks help to motivate the following definition.

Path and Path Connected

4.4.1 Definition. Let X be a space and I = [0, 1]. A path in X from x to y is a continuous map f: I→ X such that f(0) = x and f(1) = y. A space X is path connected if every two points of X can be joined by a path in X. f(0 is called the initial point of the path and f(1) is called the terminal point of the path f.

4.4.2 Definition: Let A be a non-empty subset of a space X is path connected if the topological Space A in the relative topology is path connected.

4.4.3 Theorem: Let X be a path connected space; then X is connected.

Proof. Suppose that X is not connected; then X has a separation. That is, X = A ∪ B. Take x ∈ A, y ∈ B. Let f: I→X be any path in X. Since f is continuous, f(I) is connected in X (by theorem 4.2.7); thus, f(I) lies entirely in A or B.

Hence, there is no path joining x to y contradicting the assumption that X is path connected. Thus, X is connected.

Remark 4.8.1 The converse of Theorem 4.3.3 is false. That is, a space may be connected and not path connected. The following is a counter-example to the converse of Theorem 4.4.3. The reader should look for other such examples.

A Space that is Connected but not Path Connected

Let $Z = \{(0, y): -1 \leq y \leq 1\} \cup \{(x, y): 0 < x \leq 1, y = \cos(\pi/x)\}$. It is no possible, picture this space completely, as x approach 0 the graph oscillation of the graph $y = \cos(\pi/x)$ becomes more and more rapid.

First let us show that Z is connected. Let us decompose Z into two sunsets z_1 and Z_2 where $Z_1 = \{(0, y): -1 \leq y \leq 1\}$, on the Y-axis, and $Z_2 = \{(x, y): 0 < x \leq 1, y = \cos(\pi/x)\}$. The function $F(t) = (t, \cos(\pi/t))$ defines a continuous function of the connected set $(0, 1]$ onto Z_2, hence Z_2 is connected. Now we must show that $Z = Z_1 \cup Z_2$. To prove this take $(0, y) \in Z_1$ since there are points of Z_2 arbitrarily close to Z_1 we have for $\varepsilon > 0$ given. There is an even integer N sufficiently large so that

$(1/N) > \varepsilon$. Now we have that

$\cos(\pi/(1/N)) = 1$ and $\cos(\pi/(1/N) + 1)) = -1$, hence by the Intermediate Value Theorem there is a number $t \in [1/(n + 1), 1/n]$ so that $\cos(\pi/t) = y$. Thus the point $(t, \cos(\pi/t))$ is in Z_2 and its distance from $(0, y)$ is less than ε. Thus $Z_1 \subset$ closure of Z_2 and we have $Z_2 = Z$.

We now show that z is not path connected. Assume that Z is path connected, then given $p = (0, y) \in Z_1$ and the point $(1, -1) \in Z_2$ then we have a(t), $F_2(t)$) path $F[0, 1] \to Z$ such that $F(0) = (0, y)$ and $F(1) = (1, -1)$. At this point let $F(t) = (F_1(t), F_2(t))$. Then F_1 and F_2 are continuous functions and $F_1(0) = 0$, $F_2(1) = 1$. The set $U = F_1^{-1}(\{0\})$ is closed and bounded Subset of real numbers and thus contains a least upper bound t^*. Since $F_1(1) \neq 0$, $t^* > 1$. For each value of t such that $t^* < t \leq 1$ we have $F_1(t) > 0$, hence $F(t) \in Z_2$ of $F(t) = (F_1(t), \cos(\pi/F_1(t))$ or $F_2(t) = \cos(\pi/F_1(t))$. Let $\varepsilon \leq 1$; we have for each $\delta > 0$, $t^* + \delta \leq 1$, there is a t such that $|t^* - t| < \delta$ whereas $|F_2(t^*) - F_2(t)| \geq \varepsilon$. First $F_1(t^* + \delta) > 0$, hence there is an even integer N sufficiently large so that $F_1(t^*) < 1/(N + 1) < 1/N < F_1(t^* + \delta)$. By the Intermediate Value Theorem, we have $r, s \in [t^*, t^* + \delta]$ such that $F_1(r) = 1/(N + 1)$, $F_1(s) = 1/N$. Since $r, s > t^*$ we have that $F_2(r) = \cos(\pi/F_1(r)) = \cos(N + 1)\pi = -1$, $F_2(s) = \cos(\pi/F_1(s)) = \cos N\pi = 1$. Thus, if $F_2(t^*) \geq 0$,

$|F_2(t^*) - F_2(r)| \geq 1 \geq \varepsilon$, whereas $F_2(t^*) \leq 0$, $|F_2(t^*) - F_2(s)| \geq 1 \geq \varepsilon$, contradicting the fact that F_2 is continuous at t^*. Thus we have shown that no such path F exist and space Z is not path connected.

4.4.3 **Theorem:** Let Y be a space and X be a path connected space with a continuous map $f:X \to Y$; then $f(X)$ is path connected in Y.

Proof. (Exercise for the reader).

4.4.4 Definition: A space X is said to be locally path connected at p if for every neighborhood U of p, there is a path-connected neighborhood V of p contained in U. If X is locally path connected at each of its points, then it is said to be locally path connected.

4.4.5 Definition: Let X be a space and $s \in X$, then β is a path component for s in X if and only if for any t in X with a path from s to t then t is in β.

4.4.6 Theorem: The path components of a space X are path connected disjoint subsets of X whose union is X, such that each path-connected subset of X intersects only one of these components.

Proof: Exercise for the reader.

4.4.7 Theorem: If a space X is locally path connected, then for every open set U of X, each path-component of U is open in X.

Proof: Exercise for the reader.

4.4.8 Theorem: Let X be a space X and for every open set U of X, each path-component of U is open in X, then X is locally path-connected.

Proof: Exercise for the reader.

4.4.9 Theorem: If X is a space, each path component of X lies is a component of X.

Proof: Exercise for the reader.

4.4.10 Theorem: If X is a locally connected space, then the components and path components are the same.

Proof: Exercise for the reader.

4.5.0 Compact Spaces

Compactness, like path connectedness and connectedness is a "global" property; that is, it depends on the nature of the space.

In the study of calculus, there are three basic theorems about continuous functions, and on these theorems the rest of calculus depends. These theorems are the intermediate value theorem, the maximum value theorem and the uniform continuity theorem.

The property of the space [a, b] on which the intermediate value theorem depends is the property of connectedness, and the property on which the other two depend is the property of compactness.

As these three theorems are fundamental for the theory of calculus, so are the notions of connectedness and compactness fundamental in higher analysis, geometry, and topology.

Covering

4.5.1 Definition. Let X be a set, $Y \subset X$ and let $S = \{A_i : i \in J$, and J an indexing set$\}$ be a family of subsets of X; then $\{A_i\}$ is a cover for Y provided $Y \subset$

$\cup_{i\in J}A_i$. If $Y = \{B_i\colon j\in I\} \subset B$, then T is a subcover for Y, provided $Y \subset \cup_{j\in I}B_j$.

4.5.2 **Definition:** A collection β of subsets of X is said to Satisfy the finite intersection property (FIP) If for every subcollection $\{A_1, A_2, \ldots A_n\}$ of β, then $\cap_{k=1}^{n} A_k \neq \varnothing$

4.5.3 **Theorem:** Let X be a space; if for each family of closed sets $\{A_i\colon i \in J\}$ of X, $\cap_{i\in J}A_i = \varnothing$; then there is a set $I \subset J$, I finite so that $\cap_{i\in I}A_i = \varnothing$, then X is compact.

Proof. Suppose each family of closed sets $\{A_i\colon i \in J\}$, $\cap_{i\in J}A_i = \varnothing$ there is a finite subfamily $\{A_i\colon i \in I\}$ with $\cap_{i\in I}A_i = \varnothing$. Let $\{U_i\colon i \in K\}$ be an open cover for X; then $X \setminus U_i$ is closed for each $i \in K$, but $\varnothing = X \setminus \cup_{i\in K}U_i = \cap_{i\in K}(X \setminus U_i)$. Thus, there is a finite set $I \subset K$ and $\cap_{i\in I}(X \setminus U_i) = \varnothing$, but $X \setminus \cup_{i\in I} = \cap_{i\in I}(X \setminus U_i) = \varnothing$. Hence, $X = \cup_{i\in I}U_i$ and $\{U_i\colon i \in I\}$, is a finite subcover for X. Thus X is compact.

4.5.4 Theorem: Let X be a space; then X is compact and for each family of closed sets $\{A_i: i \in J\}$ of X, $\cap_{i\in J}A_i = \emptyset$; then there is a set $I \subset J$, I finite so that $\cap_{i\in I}A_i = \emptyset$.

Proof. Suppose X is compact, and let the set $\{A_i: i \in J\}$ be any family of closed sets in X so that $\cap_{i\in J}A_i = \emptyset$. Take $U_i = X \setminus A_i$; then U_i is open and $\cup_{i\in J}U_i = \cup_{i\in J}(X \setminus A_i) = X \setminus \cap_{i\in J}A_i = X \setminus \emptyset = X$ and $\{U_i: i \in J\}$ is an open cover for X, X compact; hence, there is a set $I \subset J$, I finite with $X = \cup_{i\in I}U_i = \cup_{i\in I}(X \setminus A_i) = X \setminus \cap_{i\in I}A_i$. Thus, $\cap_{i\in I}A_i = \emptyset$.

4.5.5 Theorem: Let X be a space; if for each family of closed sets $\{A_i: i \in J\}$ of X, $\cap_{i\in J}A_i \neq \emptyset$; then there is a set $I \subset J$, I finite so that $\cap_{i\in I}A_i \neq \emptyset$ (has FIP), then X is compact.

Proof: Exercise for the reader.

4.5.6 Theorem: Let X be a space; then X is compact and for each family of closed sets $\{A_i: i \in J\}$ of X,

$\cap_{i \in J} A_i \neq \emptyset$; then there is a set $I \subset J$, I finite so that $\cap_{i \in I} A_i \neq \emptyset$ (has FIP)

. Proof: Exercise for the reader.

4.5.7 **Theorem:** Let A be a compact subset of a Hausdoff Space X, then A is closed.

Proof. Take $p \in X \setminus A$, then $p \in A$. Now for each $x \in A$, there exist U_x, V_x with $p \in V_x$ and $x \in U_x$ such that $V_x \cap U_x = \emptyset$, since X is Hausdoff. Now $A \subset \cup_{x \in A} U_x$, A compact we have that $A \subset \cup^n{}_{i=1} U_{xi} x_{i \in A}$ and $p \in \cap^n{}_{i=1} V_{xi}$, $x_i \in A$, take $V = \cap^n{}_{i=1} V_{xi}$ and ${}_iU = \cup^n{}_{i=1} U_{xi}$ for $x_i \in A$; then $p \in V \subset X \setminus A$ and V is open. Hence $X \setminus A$ is open and, thus, A is closed.

4.5.8 **Theorem:** Let A be a closed subset of a compact space X; then A is compact.

Proof. (Exercise for the reader).

4.5.9 **Theorem:** Let X be a compact space and $f(X) \to Y$ be continuous; then f(X) is compact.

Proof. (Exercise for the reader).

4.5.10 Theorem: Let $\pi_{i \in J} X_i$ be compact with the product topology; then X_i is compact for each $i \in J$.

Proof. (Exercise for the reader).

4.5.11 Theorem: Let R be the set of real numbers with the usual topology; then we have the following:

(a) [0, 1] is compact in R

(b) (0, 1), (0, 1] and [0, 1) are not compact in R.

Proof. (a) Let $\{U_i\}$ $i \in J$ be a covering for [0, 1] and let c = sup {x: [0, x] can be covered by finitely many V_i}. If $c \geq 1$, the proof is complete. Note that $c \in U_k$ for some $k \in J$; thus, there is a $B(c; t) \subset U_k$. Observe that for this positive real number t, [0, c-t/2] can be covered by finitely many of the U_i's, say U_{p1}, $U_{p2}, \ldots U_{pn}$. Now [0, c + t/2] can be covered by U_{p1}, $U_{p2}, \ldots U_{pn}$, U_k, which is finitely many of the U_{pi}'s, contradiction. Hence, $c \geq 1$ and [0, 1] is compact.

(b) Let $S = \{U_n\colon\ U_n = (0, 1 - 1/2n),\ n = 1, 2, \ldots\}$ be covering for (0,1) with no finite subcover and $S' = \{U_n\colon\ U_n = (1/2n, 2)\colon 1, 2, \ldots\}$ be a covering for (0, 1] with no finite subcover.

4.5.12 Corollary: Let R be the real numbers; then each closed interval [a, b] is compact.

Proof. (Exercise for the reader).

4.5.13 Theorem: Let X be a Hausdoff Space, and A, B, are a pair of disjoint compact subsets. Then there exists a pair of disjoint open sets U and V such that $A \subset U$ and $B \subset V$.

Proof. Since A and B are compact by Theorem 4.4.6, A and B are closed. We first show that for $p \in B$, there exist open sets W_p and W_p' with $p \in W$ and $A \subset W'$ such that $W_p \cap W_p' = \emptyset$. $p \in B$; then $p \in X \setminus A$ and $X \setminus A$ is open. Since X is a Hausdoff, for each $x \in A$ there are open sets V_x, U_x with $p \in V_x$ and $x \in U_x$, $U_x \cap V_x = \emptyset$. Now $\{U_x\}$, $x \in A$ is an open cover for A, A compact we have a finite subcover $\{U_x\}^n_{i=1}$ take $W_p = {}^{n\rightarrow}\cap_{i\in 1} V_{xi}$, $W_p' = {}^{n\rightarrow}\cup_{i\in 1} U_{xi}$. Then $p \in W_p$ and $A \subset W_p'$, $W_p \cap W_p' = \emptyset$. Now we may do this for each $p \in B$ and take $V = \cup_{p\in B} W_p$ and $U = \cap_{p\in B} W_p'$. Thus, $A \subset U$, $B \subset$ and $U \cap V = \emptyset$.

4.5.14 Theorem: The product of any finite collection of compact is compact.

Proof: We give a proof for two compact spaces have a product that is compact. The theorem will follow by induction for any finite product.

Firstly, suppose spaces X and Y, with Y compact. Suppose x_0 is a point in X, and N is an open set of X x Y containing the "slice" x_0 x Y of X x Y. Now let us cover x_0 x Y by basis elements U x V (for the topology of X x Y) lying in N. Since the space x_0 x Y is compact and homeomorphic to Y. Thus we can cover x_0 x Y by finitely many such basis elements

$$U_1 \times V_1, \ldots \ldots U_n \times V_n.$$

Also, $U_i \times V_i \cap x_0 \times Y \neq \Phi$ for $1 \leq i \leq n$ otherwise it is not needed to cover x_0 x Y. Let $\{U_i \times V_i\}_{i=1}^n$ cover the slice x_0 x Y

$$\text{Define } W = U_1 \cap U_2 \cap \ldots \ldots \cap U_n.$$

Now W is open and $x_0 \in W$. Hence $\{U_i \times V_i\}_{i=1}^n$ cover the slice W x Y.

For the point $x \times y \in W \times Y$. Consider the point $x_0 \times y \in x_0 \times Y$, now $x_0 \times y \in \{U_i \times V_i\}_{i=1}^n$ since $x \in W$, $x \in U_i$ for each $1 \leq i \leq n$, thus $x \times y \in \{U_i \times V_i\}_{i=1}^n$.

Since all the sets U_i x V_i le in N and they cover W x Y, the tube W x Y lie in N.

Secondly, suppose X and Y are both compact. Let α be an open covering for X x Y. Given $x_0 \in X$, the slice x_0 x Y is compact and has a finite subcover $U_1, \ldots \ldots U_m$ of α. Now by part 1 we have $N = \cup_{I=1}^{m} A_i$ contains the tube W x Y.

Thus for each x in X, we can choose a neighborhood W_x of x such that the tube W_x x Y can be covered by finitely many elements of α. Then the collection of all neighborhoods W_x is an open covering of X, by the compactness of X there is a finite sub collection

$$\{W_1, \ldots \ldots W_k\}$$

covering X. Now the union of all of the tubes

$\{W_1$ x Y, w_2 x Y, $\ldots \ldots W_k$ x Y$\}$ will cover all of X x Y

so X x Y is compact.

Exercise Set 4-5

1. Is it true that if X is connected and $\emptyset \neq A \subset X$, then Bdry (A) $\neq \emptyset$?

2. A space is totally disconnected if and only if the only connected subsets are one-point sets. Is every finite T_2-space totally disconnected?

3. Let f: [0, 1] be continuous show that for some $x \in [0, 1]$, $f(x) = x$. In such a case the function f is said to have a fixed point. Hint: Define $g(x) = x-f(x)/|x+f(x)|$ where g: $[0, 1]\to[0, 1]$).

4.6.0 Locally Compact

Many of the spaces occurring in calculus are not compact, but have a local version of compactness.

Locally Compact

4.6.1 Definition: A space X is locally compact at $p \in X$ if and only if there is a compact set $A \subset X$, and A contains a neighborhood of p.

Example 4.6.1 The real number R is locally compact, since for each $p \in R$, $T = [p - t, p - t]$, $t > 0$ is compact and $(p - t, p + t) \subset T$.

4.6.2 **Theorem:** Let X be a Hausdoff space that is locally compact; then for each $p \in X$, there is an open set U with $p \in U$ and U^c compact.

Proof. (Exercise for the reader).

4.6.3 **Theorem:** Let X be a locally compact, Hausdoff space, with U open and $p \in U \in X$; then there is an open set V with V^c compact and $p \in V \subset V^c \subset U$.

Proof. (Exercise for the reader).

4.6.4 **Theorem:** Let B be a compact subset of a locally compact, Hausdoff Space, with U open and $B \subset U$; then there exists a relative compact neighborhood V (V a neighborhood with V^c compact.

Proof. (Exercise for the reader).

Made in the USA
Monee, IL
08 November 2026

67382829R00096